Fishing Florida's Top 10 Bass Lakes

Volume I: Lake Tarpon

Photo Descriptions

Front Cover Photos

Upper left:

Larry Barker with a Lake Tarpon trophy bass

Upper right:

Pat Fisher with a 3 pound Lake Tarpon bass — her first bass ever!

Lower right:

Lenny Crispino with a Lake Tarpon beauty

Photos in Text

Chapter 1 cover page:

Larry and Lenny weigh in the limit of 5 bass in a Lake Tarpon tournament

Chapter 2 cover page:

Jerry and Jody Kennedy display 2 big bass that helped them win 1st place in a Lake Tarpon tournament

Chapter 3 cover page:

Melissa Barker shows off a nice cold weather Lake Tarpon bass

Chapter 4 cover page:

Lenny holds his son John's trophy bass

Chapter 5 cover page:

Riley Witt, in 1961, holding his Lake Tarpon record 19 lb bass

Chapter 6 cover page:

Catherine and Jay Weil having a good fishing day on Lake Tarpon while vacationing in the Tampa Bay area

Appendices cover page:

Bob Barker and Dad with Bob's trophy bass

Fishing Florida's Top 10 Bass Lakes

Volume I: Lake Tarpon

Larry Barker, Ph.D.
and
Capt. Lenny Crispino

Dan Richardson, Contributing Editor

BEI Outdoors Publishing

Printed in Victoria, Canada

BEI Outdoors Publishing
Book Division
30617 U.S. Highway 19 N., Suite 630
Palm Harbour, Fl 34684

National Library of Canada Cataloguing in Publication Data

BEI Outdoors Publishing
Lake Tarpon / Larry Barker and Larry Crispino.
(Fishing Florida's top 10 bass lakes ; v. 1)
ISBN 1-55395-244-8
I. Title. II. Series.
SH483.B37 2003 799.1'773'09759 C2002-906102-4

TRAFFORD

This book was published *on-demand* in cooperation with Trafford Publishing.
On-demand publishing is a unique process and service of making a book available for retail sale to the public taking advantage of on-demand manufacturing and Internet marketing. **On-demand publishing** includes promotions, retail sales, manufacturing, order fulfilment, accounting and collecting royalties on behalf of the author.

Suite 6E, 2333 Government St., Victoria, B.C. V8T 4P4, CANADA
Phone 250-383-6864 Toll-free 1-888-232-4444 (Canada & US)
Fax 250-383-6804 E-mail sales@trafford.com
Web site www.trafford.com TRAFFORD PUBLISHING IS A DIVISION OF TRAFFORD HOLDINGS LTD.
Trafford Catalogue #02-0958 www.trafford.com/robots/02-0958 html

10 9 8 7 6 5 4 3 2 1

Contents

Chapter 2

Acknowledgements

We want to thank several folks for helping make this book possible. First of all, thanks to Melissa Barker for serving as Senior Editor for the project. Without her participation this book would never have gotten off the ground (or off the water, as the case may be).

Thanks also to Karen Williams Seel, Vice-Chairman of the Pinellas County Commissioners for her time and interest in the project, and to Jacob Stowers, Assistant Pinellas County Administrator, Donald Hicks and Andrew Squires of the Pinellas County Department of Environmental Management for their time and input concerning the future of Lake Tarpon.

Thanks to Greg Whalley , Jody Kennedy and Jerry Kennedy for their encouragement and willingness to share fishing secrets; and, finally, to Dan Richardson, our mutual friend and professional tournament fisherman, for his significant contributions to the book and for serving in the role of contributing editor.

Preface

In April, 2001, Larry and Melissa Barker bought a townhouse facing the Lake Tarpon Outfall Canal. It was a rather sudden decision, given that they were renting a condo at Indian Rocks Beach at the time, and had just decided to look at properties "on a whim." But, fate is much stronger than will, and the townhouse just sort of appeared out of the blue. They fell in love with it and the beautiful view. Neither had ever seen Lake Tarpon before the day they made their offer to buy the property. What a great surprise!

Larry began to do some inquiring and net surfing, and learned that Lake Tarpon was one of the state's premier bass lakes. One brochure from the Chamber of Commerce even referred to Lake Tarpon as " The Jewel of Pinellas County." Being an avid fisherman and outdoors writer, he scheduled a guide trip with Captain Lenny Crispino, reputed to be one of the lake's best guides.

The day for the trip came, and the two fishermen met for the first time at Tarpon Tom's Bait and Tackle early one cold April morning. Lenny, to Larry's pleasant surprise, was not only a great guide, but he was courteous, patient and friendly. (As some readers will attest, these qualities are not always associated with Florida guides).

On that first trip, Larry learned that Lenny had moved to Florida from New York City in the 80's, in large part to pursue his favorite pastime – fishing. When he first moved to the Tampa Bay area, fishing was just a sideline. However, after a few years, Lenny's ability to consistently find quality fish and help his clients catch trophy bass allowed him to make the move into full time professional guiding. In the late 1990's Lenny purchased Tarpon Tom's Bait and Tackle, a business that has provided Lake Tarpon and saltwater anglers with bait, tackle, guide service and advice for over 20 years.

Fast forward to the present. That first trip was just a "teaser." Larry and Lenny have fished together dozens of times since then, including fishing as partners in professional bass tournaments, and have caught some pretty impressive numbers of trophy bass. One incident worth noting is that on their first outing, Lenny mentioned that he had not been "skunked" as a guide for the past 7 ½ years. Larry brought that streak to an end several months later when they had only four bites all day and Larry was unable to get a hook to stick in even one. He felt terrible, but Lenny laughed it off and said that it was time to start a new streak. That may be the day when the fishing relationship turned to a friendship.

The idea of writing this book came soon thereafter. Larry realized that there was relatively little information available about fishing Lake Tarpon, and that Lenny was one of its premier professional guides and bass experts. Both were experienced outdoors writers and tournament fishermen, so they decided to team up and prepare this book to help locals and visitors alike catch more and better quality fish in Lake Tarpon.

We hope you like the results of our research and experience. We'd love to hear from readers with any comments, criticisms or suggestions.

Tight Lines!

Larry Barker and Lenny Crispino

Florida's Top Fishing Lakes

The Florida Fish and Game Commission has compiled a list of the top 10 lakes in the state for catching bass and crappie, and the top 15 for catching panfish. (The Commission has also named the top 10 catfish holes – some of which are in rivers). The lists are created on the basis of creel surveys, angler interviews and reports from fisheries experts. Some lakes are noted for producing record numbers and/or sizes of only one species – others are noted for exceptional fishing for multiple species. Given that Florida boasts over 7,500 lakes (not to mention about 12,000 miles of rivers and streams), it can be tough to decide which of the state's lakes to fish. The books in this series are designed to help anglers make educated choices about where to fish – in order to catch the most and biggest fish in a limited period of fishing time. The state has done the tough job of picking the best lakes; we now want to add to that effort by giving anglers high quality information about each of the top lakes.

The first ten books in the series will focus on the Fish and Game Commission's list of top ten Black Bass lakes in Florida. The lakes are not rank ordered, so their volume number does not indicate any relative place within the top ten. Our books also include information about the other fish available in each of the top lakes. As noted above, several of the lakes are listed in the top ten for more than one species.

Florida's Top Fishing Lakes as Compiled by the Florida Fish and Game Commission

The following lists of Top 10 Bass and Crappie Lakes and 15 Panfish Lakes are presented in random order. The lists below reflect selections for 2002. However, the lists can change from year to year, so interested anglers should consult the Florida Fish and Game Commission annually for updated lists.

Top 10 Black Bass Lakes

Lake	Location
Lake Tarpon	Palm Harbor/Tarpon Springs
Lake George	Deland/Ocala
Stick Marsh/Farm 13 Reservoir	Vero Beach
Lake Kissimmee	Lake Wales
West Lake Tohopekaliga	Kissimmee
Rodman Reservoir	Palatka
Lake Weohyakapka (Lake Walk - in – Water)	Lake Wales
Lake Istokpoga	Sebring
Everglades Water Conservation Areas 2 & 3	Southern Florida
Lake Okeechobee	South Central Florida

Top 10 Speck (Black Crappie) Lakes

Lake	Location
Lake Kenansville	Vero Beach
Lake Marian	Haines City
Lake Trafford	Fort Meyers
Lake Harris	Leesburg
Lake Monroe	Sanford
Lake Talquin	Tallahassee
Lake Okeechobee	South Central Florida
Tenoroc Fish Management Area	Lakeland
Lake Woodruff	DeLeon Springs
Lake Weir	Ocala

Top 15 Panfish Lakes

Lake	Location
Lake Monroe	Sanford
Lake Kissimmee	Lake Wales
Lake Okeechobee	South Central Florida
Lake Kenansville	Kenansville
Lake Panasoffkee	Sumter County
Lake Walk-in-Water	Lake Wales
Johns Lake	Clermont
Lake Talquin	Tallahassee
The Fab Five (five small lakes near Orlando: Starke, Turkey, Kirkman Pond, Cane-Marsha Park, Clear Lake)	Orlando
Tenoroc Fish Management Area	Lakeland
Lake Harris	Leesburg
Lake Marian	Osceola County
Lake Istokpoga	Sebring
Lake Jessup	Sanford
Choctawatchee River	Florida Panhandle

Introduction

Lake Tarpon is one of west-central Florida's best kept secrets. Although as we previously noted, it was named one of the Top 10 Black Bass Lakes in Florida by the Florida Fish and Game Commission, fishing pressure on Lake Tarpon has been very light in recent years. One reason is that it is located near prime saltwater areas such as Tampa Bay and the Gulf of Mexico – and many local and visiting anglers spend their fishing time going after saltwater species such as snook, cobia, kingfish, grouper, and tarpon.

Another reason the lake isn't fished heavily may be its location. Its 2500 plus acres are nestled in an area of Northern Pinellas County that is primarily residential and commercial. If you didn't know that the lake was there, you'd probably never find it. U.S. Hwy. 19 N borders the lake on the West – but almost no open water can be viewed from the road. There are only three easy public access areas and two of these are through State Parks. Most state maps show Lake Tarpon but many regional and national maps don't even show it.

This book is written for local and visiting fishermen (and women)* who want to catch more and bigger fish in Lake Tarpon. Our focus is primarily on black bass fishing, with an emphasis on trophy bass. However, it also includes fishing tips and specific locations for the lake's other species such as bream (bluegill), shell-cracker, specks (crappie), catfish, gar and bowfin. All of these species are abundant in Lake Tarpon and usually are willing to bite year 'round.

---

*We use the term "fisherman" in this guide to designate the person who fishes. Being aware that many women enjoy fishing and are highly skilled, we don't want to slight them in the least. However, we feel that the term "fisherperson," although more politically correct, is cumbersome and sounds a bit contrived.

Chapter One

About Lake Tarpon

Lake Tarpon

Pinellas County, Florida

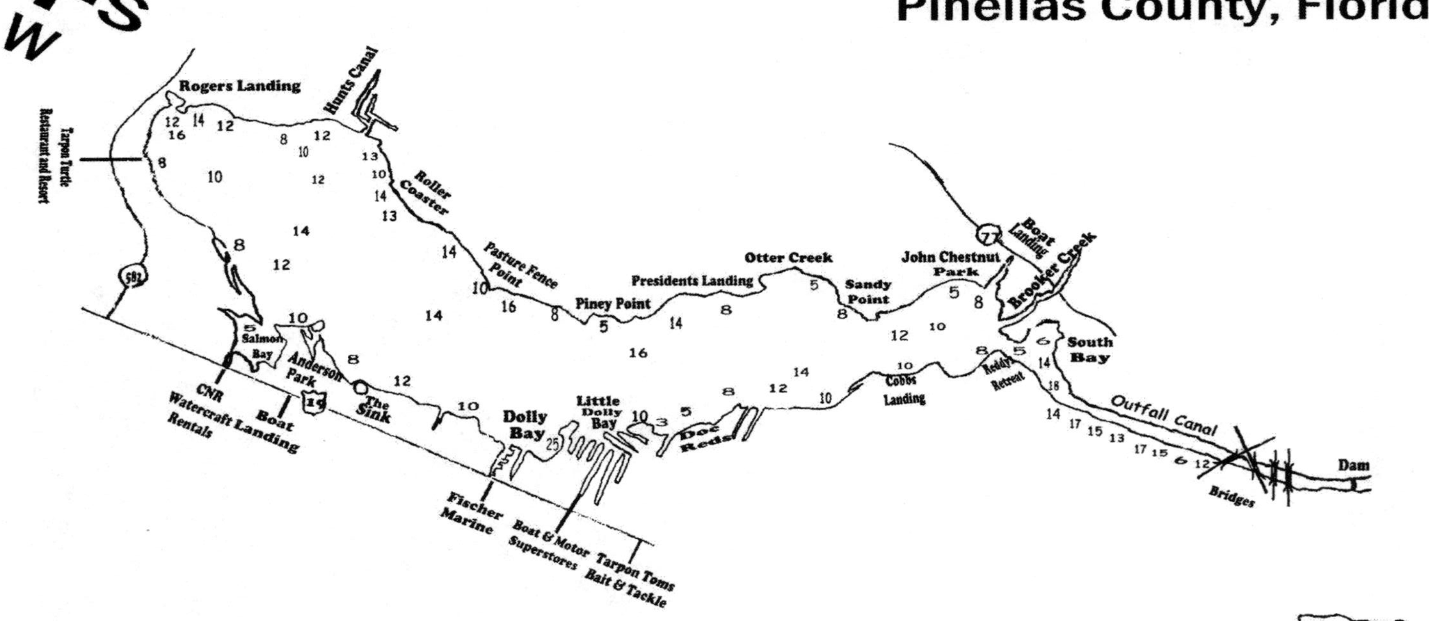

0 1/2 1

Miles

(Not for navigation)

L & L FISIHING MAPS

Chapter One

About Lake Tarpon

Location

Lake Tarpon is a 2,534-acre Fish Management Area* located in Pinellas County, Florida, near the Tampa/St. Petersburg/Clearwater metro area. The upper portion of the lake is in the city of Tarpon Springs. The lake is approximately 17 miles northwest of the Tampa International Airport. It is about 10 miles north of the St. Petersburg-Clearwater Airport and approximately 15 miles west of the Tampa Executive Airport (servicing corporate and private aircraft). Directions to Lake Tarpon access points can be found on several web sites, including tampabay.com, TBO.com, and tarpon-toms.com. For directions by phone, call Tarpon Tom's Bait and Tackle at (727) 938-2379.

Public access to the lake is available through two county parks. Anderson Park is on the west shore, off of U.S. Highway 19 N. The entrance is about .9 mile south of the intersection of US 19 and County Road (CR) 582 (Tarpon Avenue) in Tarpon Springs

---

*A fish management area (FMA) is a body of water established for the management of freshwater fish as a cooperative effort between the Florida Fish and Conservation Commission and local counties. For the most current regulations regarding Lake Tarpon, consult the Florida Freshwater Sportfishing Guide and Regulations Summary. (available on line at http://floridafisheries.com/fma)

John Chesnut Park is on the east side of the lake, off of C.R. 611 (East Lake Road). To reach Chesnut Park from U.S. 19, take either C.R. 584 (Tampa Road) or C.R. 586 (Curlew Road) east to McMullen-Booth Road. Then, turn north on McMullen-Booth Road and go about 2 miles from C.R. 584 or approximately 2.1 miles from C.R. 586 to East Lake Road. Turn left on East Lake Road and follow it to the park. Both parks open at 7 a.m. and close at dark year round. (*Note: If you use park boat ramps and leave your boat trailer in the parking lot, remember to check the sign at the entrance telling what time the park closes that day).*

Rogers Landing (formerly called Noell Fish Camp), off of Keystone Road (C.R. 582) is another lake access point. It is located in North Cove. There you can use the boat ramp for a nominal fee. The boat ramp at Rogers is open 24 hours a day, 7 days a week.

More Information About the Parks

Anderson Park (39699 US Hwy. 19 N., Tarpon Springs, FL 34689; Ph: 727-943-4085) boasts 128 acres and lies along the north shore of Lake Tarpon. It was dedicated in 1966 and offers group picnicking facilities, play equipment and a boat launching facility. The park is heavily wooded and offers beautiful views of Salmon Bay (adjacent to the park) and Lake Tarpon. Anderson Park also has a 478 ft nature trail among cypress trees along the Salmon Bay shoreline, plus an elevated boardwalk. Fishing from the bank is allowed within the park.

John Chestnut, Sr. Park (2200 East Lake Rd., Palm Harbor, FL 34685; Ph: 727-669-1951) covers 255 acres and was dedicated in 1979. Located on the southeast corner of Lake Tarpon, its facilities were designed to complement the natural beauty of the park. Facilities include over 6800 feet of nature trails, a 30 foot high observation tower, open play areas, picnic shelters, restrooms, a ballfield and playground equipment. The park also has several boat ramps that provide access to the lake by way of a 750 foot long channel.

A Brief History

Lake Tarpon, with a surface area of about four square miles, is the largest freshwater body of water in Pinellas County. The lake is approximately 6 miles long and varies in width from about ½ to 1 mile. The average depth of the lake is around 7.5 feet, with the maximum depth about 25 feet.

Formerly called Lake Butler, Lake Tarpon has been used for boating, fishing and swimming by locals and visitors since the early 1920's. (The name was changed from Lake Butler to Lake Tarpon because there was another lake named Lake Butler near Orlando). Lake Tarpon was used as a water supply in Pinellas County over a four-year period between 1926 and 1930. The lake water then became too salty for drinking because of water flowing into the lake through a sinkhole, called Lake Tarpon Sink.

Lake Tarpon Sink, 118 feet deep, is connected underground with the saltwater Spring Bayou in Tarpon Springs, and the water from the sinkhole initially flowed into the lake at high tide. Until the late 1960's, the sinkhole acted as both an outflow and inflow for Lake Tarpon, depending on the tides and the water level of the lake. In order to maintain Lake Tarpon as a freshwater body of water, the District constructed an earthen berm around Lake Tarpon Sink in 1969. This effectively contained all salt water in the sink, and it could no longer flow into the lake. Two years earlier, as part of a larger project, the United States Army Corps of Engineers had designed and constructed an Outfall Canal at the southernmost end of the lake to provide flood control for Lake Tarpon.

The **Outfall Canal** is not only useful as a flood control device, but provides an outstanding environment for both panfish and trophy bass. The Canal is about 3 1/5 miles in length, but only about 1 mile of the Canal connects with Lake Tarpon. A dam serves as a barrier between the saltwater and freshwater areas. Beyond the dam, the saltwater segment connects with Tampa Bay. With some effort, a small boat can be dragged across the dam from saltwater to freshwater and vice versa.

The dam helps keep the lake from overflowing its shores, and also allows officials to regulate water levels to control lake vegetation. Once in the 1990's, the lake was inadvertently lowered two feet as a result of human error in closing the dam gates. However, because **Brooker Creek** is such a robust watershed, it filled the lake back to its original level within four days!

Brooker Creek runs approximately 15 miles and drains about 42 miles of Pinellas County and northwest Hillsborough County before entering Lake Tarpon at its lower southeastern corner, less than 3000 feet upstream of the Outfall Canal. The headwaters of Brooker Creek consist of fourteen channels that eventually converge and form the main channel, which flows into Lake Tarpon. The area where Brooker Creek empties into Lake Tarpon is an outstanding fishing area, providing all the right conditions for trophy bass and panfish alike.

The Future of Lake Tarpon

Although the history of Lake Tarpon as a fishing management area is important to anglers, the future of the lake for sport fishing is probably more important. As part of our research for this guide, we interviewed representatives from several groups overseeing the quality and future of the lake. These included individuals from the Pinellas County Board of County Commissioners, the Pinellas County Department of Environmental Management - Water Resources Management Section, and the Southwest Florida Water Management District.

Based on our interviews, it is safe to say that Lake Tarpon promises to continue as a high quality fish management area in the immediate future. Although the present quality of the lake is somewhat lower than it was two decades ago (based on Department of Environmental Management criteria), it is still considered a productive fishery.

Plans are in place to restore and maintain balance among fish and vegetation in Lake Tarpon on a continuing basis. Several projects and assessments are currently underway to deal with such lake quality issues as phosphorous inactivation and precipitation (lake alum treatments), sediment removal (dredging) , flushing and dilution (lake drawdowns) and biomass removal (harvesting aquatic vegetation). Budgets are tight, but adequate funds appear to be earmarked for the most critical of these projects.

Interest within local, state and federal government agencies in preservation of lake quality is evident, as is that of local fishermen, guides and lake related businesspersons.

Perhaps the biggest threat to the lake's quality comes from run off of fertilizers and lawn chemicals from adjacent properties. There is little coordinated effort at present to help educate lake residents and/or regulate such run offs. Karen Williams Seel, Vice-Chairman of the Pinellas County Board of County Commissioners, at the time of this writing, is exploring possible ways to educate the public and to coordinate efforts of a variety of governmental agencies, private organizations, and individuals interested in the welfare of Lake Tarpon. Hopefully, an advisory committee will be established to address education and coordination issues concerning the lake.

Chapter Two

Lake Tarpon Fish

Chapter Two

Lake Tarpon Fish

We assume that anyone reading this guide probably already knows something about the kinds of fish found in Lake Tarpon. Therefore, we won't give a lot of detail about fish identification and classification. If you want more technical information about a given species of fish, we suggest that you check either Florida Sportsman's *Sport Fish of Florida* by Vic Dunaway (FS Books, 2000), or the Florida Fish and Wildlife Conservation Commission's website: *floridafisheries.com/fishes*. Both provide excellent references concerning habits and habitats of all the fish that swim in Lake Tarpon.

We do want to provide some information about the primary species that anglers pursue in Lake Tarpon, to help you make better informed decisions about when, where, and how to fish for each species in the lake, as well as the tackle and techniques that will be most effective. (*(Note: Since the most appropriate fishing methods and tackle are different at different times of year on the lake, seasonal changes in techniques and locations are discussed in Chapter 3, "Fishing Lake Tarpon through the Seasons").*

Largemouth Bass

Bass fishing is king on Lake Tarpon. This is not surprising since bass are, by far, the most sought-after freshwater game fish in Florida. Another reason that bass rule the lake is that they bite year 'round. Florida Fisheries biologists have observed many anglers catching 20 or more bass per day on Lake Tarpon, including an occasional 10 plus pounder! The average size of Lake Tarpon bass is 12-16 inches, but quality and trophy sized fish are present in large numbers.

One year old bass in Lake Tarpon average about seven inches in length and grow to adult size of about 10 inches in 1 to 1 ½ years. By age two, female bass tend to grow much faster than male bass, and the females also live longer. Males seldom exceed 16 inches, while females frequently grow upwards of 28 inches. Five year old females usually are about twice the weight of males. A 10 pound trophy bass will be about 10 years old!

Although there are three distinct kinds of Black Bass in Florida waters, Lake Tarpon boasts the Florida Largemouth – a species that eats more and grows faster and larger than its cousins, the Southern and Northern Largemouths.

Preferred Habitat

The bass in Lake Tarpon change preferences for habitats with each season, and with varied weather conditions. Most of the year, however, Lake Tarpon bass prefer areas where heavy patches of aquatic weeds (*particularly hydrilla)* or reeds are available. The exception to this is during the hot summer months, when, during daylight hours, the largest bass seem to prefer deep water near bottom structures and brush piles. Other favorite habitats include boat docks, eelgrass and anywhere bait fish are plentiful.

Although bass can tolerate a wide range of water clarity and bottom types, they prefer water temperatures from 65-85 degrees. It's fortunate that Lake Tarpon's temperature stays within this preferred range most of the year.

Weather Influences

Weather influences on the lake can be significant. Probably the best weather for fishing the lake is when the barometer is steady and there is cloud cover. Some anglers prefer a slowly rising or falling barometer. The period just before a cold front that disperses hot humid weather sets in often will provide excellent fishing on the lake. Depending on the fishing techniques you are using, a slight wind also can be beneficial. (*Exceptions are when fly fishing and casting light lures on bait and spin casting outfits*).

The toughest fishing, weather wise, is when there is a "bluebird" sky and the lake looks like glass; there is no wind. Under these conditions, bass can still be caught, but, given that the ability of fish to see suspicious lines, lures and objects is enhanced under these conditions, anglers must be very cautious about "spooking" fish. (*One tip, under bluebird skies, is to cut off the boat motor far away from the target fishing spot, and use the trolling motor at the slowest speed to reach the area desired. Keep boat noise to a minimum; some anglers even believe that it helps to turn off the fish finder.*)

Spawning Habits

Lake Tarpon bass usually spawn between December and May, with February and March tending to be the height of the spawning period most years. Spawning usually begins when water temperatures in the lake reach 58-65 degrees and it continues as lake temperatures climb into the 70's. Anglers can spot beds around the shallower areas in the lake (see map), usually in about 2-4 feet of water. Males build saucer-shaped nests about 3 ft. in diameter, usually in hard bottom areas found in protected locations such as canals and coves. Female bass lay their eggs (upwards of 100,000) in the bed. The eggs are then fertilized as they settle into the nest.

After spawning is completed (usually over a period from 5 to 10 days), the male bass guards the nest and eggs, and later protects the young fish (fry). The female tends to stay near the nest and remain listless for up to a day. After hatching, the fry swim in tight schools, disbanding when they reach a length of about one inch. At this point, the fry become fair game as snacks for adult bass and other species of fish and wildlife in the immediate area.

Feeding Habits

The diet of the bass changes as it grows. Young bass feed on zooplankton and small crayfish. Fingerling bass feed on insects, crayfish and small minnows. Adult bass eat almost anything! In Lake Tarpon they have been known to eat snakes, small mammals, salamanders, frogs, turtles and even birds. The lake record, which is a 19 lb bass, was caught on a live eel. However, the favorite food of bass, throughout most of the year, is other fish, particularly lake (Golden) shiners, bluegills, shad and smaller bass.

Tackle and Baits

Fishermen from "up north" usually are surprised by the heavy bass tackle recommended by local Lake Tarpon guides. Though light tackle can be fun, and even most effective in some areas of the country, it often results in heartbreak and disappointment when that 10 pound trophy bass is finally hooked, but runs unchecked into weeds or structure -- never to be brought to boat! The following tackle is recommended for each specific type of fishing:

Shiner Fishing

- 7 to 7 ½ ft medium to heavy rod with bait casing reel (clicker a must)
- 30 to 50 pound braided line with a 20 lb monofilament leader is preferred to float shiners with a bobber
- 20 to 30 pound monofilament is usually used when "live line fising." (More on this technique later)
- Hooks are usually 3/0 – 5/0
- Most good sized bobbers will work – but cork bobbers tend to cast better and blend in with the water

Baitcasting with Artificial Lures

- 7 to 7 ½ ft medium to heavy rod with precision bait casing reel
- 15 to 20 pound braided or monofilament line
- Assortment of baits including weedless plastic worms (Florida rig, Texas rig and Carolina rig); spinnerbaits, weedless spoons, topwater plugs, jerk baits, crank baits and topwater plugs

Spincasting with Artificial Lures

- 6 ½ to 7 ft medium to heavy rod with open or closed face spinning reel
- 12 to 15 pound monofilament line
- Assortment of baits including spinnerbaits, weedless spoons, crank baits, jerk baits,
- topwater plugs and swimming plugs

Fly Fishing

- 8 to 9 ft 6-8 fly rod with precision fly reel
- 6-7 weight fly line
- 6-8 lb fly tippet
- Assortment of large flies/lures including large popping bugs and big streamer flies often tied to resemble worms and eels)

Selecting Lures for Bass at Different Depths

Lure selection is based on many factors, including the angler's comfort with a lure, advice from local guides and tournament fishermen, TV commercials and print ads, just to name a few. Regardless of specific brands and types, the most effective lure selection is made with the depth of the bass in mind. The map of Lake Tarpon in this book provides a general depth guide, but your electronic fish/depth finder can also be a valuable tool. Use your depth finder to locate where fish are holding up or cruising – then use the following guide to select the appropriate lure(s):

Shallow Water (1-3 ft)

Use lures such as topwater and shallow-running plugs, weedless baits, buzz-baits, spinnerbaits, plastic baits on Florida or Carolina rigs, flippin' baits or pitchin' jigs.

Medium Water (3-8 ft)

Try diving crankbaits, lipless sinking crankbaits, buoyant plugs, Texas or Florida rigged plastic worms and unweighted worm rigs.

Deep Water (8 –20 ft)

Cast deep diving plugs and slow-rolling spinnerbaits. Try fighting heavy jigs, weighted plastic baits, jigging spoons and blade baits.

Bass usually prefer smaller and more slowly moving baits in cold water; and larger, faster moving baits in warm water. Experimenting with lure size and color is essential to find the right combination, regardless of depth of the bass.

Black Crappie

Crappies (or specks, as they are referred to locally) are plentiful in Lake Tarpon, and are highly sought after by anglers, particularly in the winter months. The Florida Black crappie is a cousin to the White crappie found in Alabama and Georgia, but is actually a separate species.

The average sized crappie in Lake Tarpon will be between 8-12 ounces. Occasionally, anglers catch specks in the one to two pound range. Though no official lake record for crappie has been kept in the past, fishermen report catching some crappies in the three pound range. Given that the current Florida state record is 3.83 lbs, it's possible that a new state record could be set by crappie fishermen in Lake Tarpon in the future.

Preferred Habitat

Crappies in Lake Tarpon are usually found in medium and deep water (4-9 feet) during the winter. Their favorite locations are in water under overhanging trees and branches, near bridge pilings, and around sunken brush piles. Don't overlook deep holes, especially in Dolly Bay!

They prefer protected areas where the current is not too swift and the water is clear. Sandy bottoms with surrounding vegetation also attract schools of specks.

Crappies tend to suspend in mid-water, so anglers have to keep experimenting to find the right depth on a given day. They prefer water in the 70-75 degree range, but will still bite when water temperatures reach the 80 degree mark.

Weather Influences

Crappies, unlike largemouth bass, don't seem to be affected too much by weather conditions. Usually, because most anglers fish for crappies during cold weather, it is usually the comfort of the fisherman, rather than the crappies' willingness to bite, that dictates whether or not to go speck fishing. High winds may be the exception. Because of the need to keep a bait drifting near a prime target area, high winds that blow floats and baits away from the area can be a nuisance that makes crappie fishing challenging.

Although crappies like quiet water, "bluebird" skies may make them extremely wary. The trick here is to anchor the boat a great distance away from the fishing spot, and cast to the target. Keeping motor and boat sounds to a minimum is important to keep from spooking the school. Night fishing is usually good for speck fishing, especially around a new moon.

Spawning Habits

Lake Tarpon crappie spawn between February and April when water temperatures reach 62-65 degrees. They tend to nest in schools. Males make circular nests over gravel or soft muddy bottoms in 3-6 feet of water, usually near weeds or vegetation.

After spawning, the males guard the eggs and fry. A half pound female will produce from 20,000 to 50,000 eggs annually. Large females can produce over 150,000 eggs per year.

Specks in the 8-12 ounce range are usually 2-3 years old. It takes two years for crappies to reach maturity. They tend to grow from 1-4 inches the first year and from 3-8 inches the second year. It takes about four years for most crappies to reach 12 inches. The oldest crappie caught in Florida, over 20 inches long, was determined to be 11 years old by the Florida Fisheries Department. Crappies tend to stay in schools (often referred to as colonies) all of their lives, from fry through adulthood.

Feeding Habits

Small crappie feed on insects and small fish. Adults mainly eat small fish. "Missouri minnows," sold in most bait shops, are a favorite. However, any sort of wild minnow or small baitfish will work.

Specks tend to bite very softly – so you have to watch your float carefully to detect bites. Because of tender mouths, light hook sets are a must, avoiding hard jerks of the rod tip. When jig fishing, the same holds true – just set the hook lightly and keep pressure on the fish. Don't horse it or rare back too hard when feeling the strike.

Tackle and Baits

Crappie fishermen are divided into two major and one minor camps regarding the best tackle for the task. One group swears by cane poles and minnows. Anglers of this persuasion anchor the boat near the target area and put cane poles out in a circular pattern around the boat. The second group prefers light and ultra-light spinning outfits, for either live bait fishing or jigging. A third, and smaller, group of fishermen prefer fly casting. Fly fishing for crappies demands a lot of patience and persistence and, more important, some good luck. The following tackle is recommended for each specific type of fishing:

Cane Pole Fishing

- 3 to 5 cane poles, each 12-15 ft in length
- 6-12 lb monofilament line is preferred – cut to the length of the cane pole
- hooks are usually long shanked, # 4 – 6
- Split shot sinkers – size varies according to desired depth and size of bait
- Sliding bobbers (usually stick type) with bobber stoppers attached to the line are probably the favorite among Lake Tarpon crappie anglers. Any kind of small float or bobber will do, providing that it will not offer much resistance when the crappie bites the minnow.
- The bobber should be set so the bait is at the target depth where the fish are holding. If this is not known, start with a depth about half the actual water depth and adjust accordingly to find the fish.

Spincasting with Artificial Lures

- 6-7 ft light rod
- 4-12 lb monofilament line
- Assortment of small leadhead jigs (chartreuse and white are always favorites) and in-line spinners, small spinnerbaits and very small swimming crankbaits.

Note: If live baits (minnows) are used with spinning gear, use the hook, split shot and bobber information above as guides to help with rigging.

Fly Fishing

- 7 to 9 ft 4-5 fly rod with reel
- 4 wt fly line (sinking preferred)
- 2 to 6 lb fly tippet
- Assortment of bugs and flies, with a good supply of small streamer flies. Some fly fishermen prefer small spinners.

Bluegill and Shellcracker

There are probably more bluegill (also called bream, sunfish, and a dozen other nicknames) per square acre in Lake Tarpon than any other species. Anglers who fish for bluegill fall into two primary groups, differentiated by their purpose for fishing. The first group views bluegill as an excellent eating fish – and one that is relatively easy to catch. Cane poles and spincasting rods and reels are the favorite tackle choices for this group.

The second group tends to view bluegill, which do belong to the same family as largemouth bass, as a sport fish, and go after them with ultra light spinning gear and fly rods. Of course, there also is a third group that views bluegill as both a sport fish and an eating fish.

Whatever the motivation to fish for bluegills, it must be "catching." Angler opinion polls conducted by the Florida Fisheries Commission show that bluegill is the most popular freshwater fish in Florida, according to the fishermen surveyed.

The name "bluegill" or "bream" is loosely given to a wide variety of fish in the sunfish family. We refer you, again, to the references provided at the beginning of this section if you are interested in learning how to identify each different species. However, it is probably useful to distinguish between bluegill and shellcracker (red eared sunfish), since both species can be caught in Lake Tarpon:

Bluegill vary a lot in color, but their distinguishing characteristic is a solid black gill flap with no color border on it. Bluegill tend to darken with age. Females usually have a light bluish tint to their scales, while males lean toward purple and rust colors. Both males and females have vertical bars on their sides. Their shape tends to be somewhat round. Bluegill average about eight ounces, but can grow to several pounds. A six inch bluegill is about two years old. The Florida record is 2.95 lbs.

Shellcrackers, on the other hand, can be easily distinguished from the bluegill, by the red spot on the end of their gill flap. The red spot is actually a border around the dark gill flap. They are multicolored, having both olive and blue sides and a bright yellow underside. They have variable dark markings, but their side stripes usually are not as pronounced as those on the bluegill. They have a more oval shape than the bluegill. Shellcrackers are larger than bluegills, in most instances averaging about twelve ounces as adults. Nine to ten inch shellcrackers are rather common in Lake Tarpon. The Florida state record is 4.86 lbs.

Preferred Habitat

Although both bluegill and shellcracker are plentiful in Lake Tarpon, their preferred habitats differ to some extent. Bluegills tend to favor the quiet, highly vegetated areas near the shoreline where they can hide from predators and find insects and worms easily. Residential canals are prime areas for bluegills. Look for stationary docks, especially those where the residents feed birds and fish.

Shellcrackers can be found near bluegill "holes," but also inhabit deeper water near shell-covered sand bars. (Shellcrackers received their name because of the grinding teeth in their throats used to crush freshwater snails, their favorite food.) Both bluegill and shellcracker change their location as water temperatures and seasons change. Therefore, the fact that you catch a bluegill in one location in the spring doesn't mean you would find fish there in the middle of summer.

Weather Influences

Although locations where bluegill and shellcracker may be found vary with the seasons, weather conditions don't greatly affect their willingness to bite. Water temperature, related to weather, may slow down or accelerate the "bite", but bluegill tend to bite almost every day of the year, and under any and all weather conditions. If the day is windy, you may find that the fish hug the shoreline a little more closely and seek structure for protection. However, once you find their hiding place, you can bet that they will take your bait readily.

Fly fishing for bluegill can be affected by the wind. Windy days make casts more difficult, but heavier popping bugs and sinking line can make fly casting easier under windy conditions.

Spawning Habits

Bluegill tend to spawn in large colonies, with their circular beds touching one another. The beds are usually in 2-5 feet of water, on sand, shell or gravel. If the bottom is soft, they may bed among plant roots. Bluegill spawn between April and October, with the peak months on Lake Tarpon being May and June when the water temperature reaches about 78-80 degrees. Female bluegill, depending on their size, lay from 2,000 to 60,000 eggs. The eggs hatch quickly after they have been fertilized – usually within a day and a half.

Shellcrackers tend to spawn earlier in the year than bluegill. Their prime spawning months in Lake Tarpon are between March and August. Their beds are usually in water from 3 to 4 feet deep, on a hard (preferably shelly) bottom, near a drop off. Spawning sites are also usually near vegetation. Female shellcrackers lay between 15,000 and 30,000 eggs per year, on average.

Feeding Habits

Bluegill are partial to red wigglers and crickets in Lake Tarpon. However, almost any insect, larvae or worm will be readily received. At certain times of year bluegill eat fish eggs, snails and even small fish. They bite at a variety of depths, depending on available cover and water clarity. In the residential canals a loaf of white bread may be all that is needed to find bluegill. Move slowly down the canal tossing whole slices of bread as you go. When you see surface strikes at the bread anchor down and have some fun.

Shellcrackers are primarily bottom feeders. They tend to forage for food during the daytime on sand bars and near vegetation. Their favorite foods are freshwater snails and clams. They also eat worms, fish eggs, small fish and a variety of insects.

Tackle and Baits

Bluegill and shellcracker can be caught on almost any tackle – from hand lines to sophisticated spinning and fly tackle. The key is to use light tackle to enjoy the fight these panfish can offer. Here are some suggestions for tackle for each specific type of

fishing:

Cane Pole Fishing

- 12 to 15 foot cane pole
- 6 lb monofilament line – cut to length of the cane pole and tied to the tip

- Number 6 or 8 short shanked hooks
- Small split shot sinker
- Small round or pencil bobber (though some cane pole fishermen prefer not to use a bobber— and just feel the bite directly)

Ultralight Spincasting

- Five to 5 ½ ft light weight graphite rod
- Ultralight spinning reel
- 2 to 6 lb monofilament line
- A variety of tiny spinnerbaits, jigs and plastic worm lures

Fly Fishing

- 7-8 ft 2-4 fly rod with reel
- 2-4 wt fly line (either sinking or floating – bug taper if using popping bugs)
- 2-4 lb tippet
- Assortment of small popping bugs, spiders, woolly buggers, and small flies. Both sinking and dry are popular – depending on time of the year and weather conditions.

Other Species

Though bass, crappie and bluegill are the three most sought after species in Lake Tarpon, anglers catch several kinds of fish with regularity. These include several varieties of catfish, gar and bowfin. We'll give some basic information about each of these fish, but will not devote as much attention to tackle and fishing methods. The reason for this is that, in Lake Tarpon, most of these fish are caught while anglers are trying to catch either bass, crappie or bluegill. The exception may be catfish, but the number of anglers who specialize in "catfishing" in Lake Tarpon is limited.

Catfish

There are several varieties of catfish to be found in Lake Tarpon. Once again, we recommend checking one of the resources noted at the beginning of this section for species identification clues.

The most common catfish in the lake is the brown bullhead. Many bluegill fishermen have been surprised when a pound plus bullhead has taken their bait—instead of the half-pound bluegill they were expecting. On light tackle or cane poles bullheads can really put up a good fight.

Occasionally a channel catfish, considerably larger than the bullhead, will take the panfish angler's bait. When this happens a real fight is in store. If the line doesn't break a lucky fisherman can land a catfish weighing several pounds. Of course, since Lake Tarpon is a fish management area, the catfish caught here are considered excellent table fare.

Bullheads average about a pound in the lake. However, some over three pounds have been reported. They are primarily bottom feeders, and tend to eat mostly in the nighttime hours. Worms, leeches, crayfish and small fish are among their favorite foods. They prefer water temperatures in the 78 to 82 degree range, but survive well in almost all water temperatures. They tend to build nests in spawning season near underwater objects such as logs or rocks. Muddy and sandy bottoms attract bullheads in quantity.

Channel catfish (also called spotted cat) can reach up to 45 lbs in weight. However, the average weight of a channel cat in Lake Tarpon is about 2 to 3 pounds. They prefer deep water with some current and a sandy or gravel bottom. Channel cats spawn when water temperatures reach 70 to 85 degrees. Like most catfish they feed best at night, using their chin barbells to locate prey. They will eat worms, insects, crayfish and small fish. Many a shiner fisherman has found a channel cat on the end of his line, when drifting for largemouth bass. Channel catfish are considered one of the best eating freshwater fish around.

Gar

Of all the varieties of garfish, the longnose and Florida gar are most common in Lake Tarpon. Few fishermen target gar as a sport fish, but they are often caught while anglers are casting artificial lures or drifting live shiners in search of largemouth bass. The truth is that gar can put up a real fight, and are exciting to catch!

Longnose Gar can reach weights in excess of 50 pounds. The average weight of those in the lake is about 5 to 7 pounds. They mainly feed on small fish and frogs, and tend to wait for their prey and strike when it comes within range. Longnose gar spawn between December and March. Their nests are usually in shallow water over weed beds or other structure. Although some anglers do not consider the longnose gar to be edible, the Seminole Indians are said to prefer gar over other fish. However, the roe from the longnose gar is poisonous to humans, animals and birds!

Florida gar are present in greater numbers than longnose gar in Lake Tarpon. They do not reach the same size, averaging only about 3 pounds. Though the state record, at present, is only 7 pounds for the Florida gar, the world record is over 21 pounds! Spawning occurs in late winter and early spring. Eggs are spread over submerged weeds and the young fish stay attached to the weeds (by an adhesive on their snout) after hatching, until they reach nearly an inch long. Like the roe of the longnose gar, that of the Florida is also poisonous.

Bowfin

Bowfin (also called mudfish and dogfish) are among the most maligned fish in Lake Tarpon! They are often caught on worms by bluegill fishermen and on shiners by bass fishermen. When they are hooked, they put up a fair fight, but anglers usually don't value them as a sport fish.

The bowfin is one of the oldest species of fish still living. Like the gars, it has an air bladder, which allows it to breath in very shallow or poorly oxygenated water. Spawning takes place in the spring near heavy weeds and grass. Young bowfin stay hidden in the vegetation for self preservation till they reach about 10 inches in length. Favorite foods of adult bowfin are fish (dead or alive), worms, frogs and crayfish. They grow very fast and often reach weights in excess of 15 pounds. Most bowfin caught in Lake Tarpon average about 5 pounds. The current state record is 19 pounds.

Chapter Three

Fishing Lake Tarpon Through the Seasons

Chapter Three

Fishing Lake Tarpon Through the Seasons

While weather conditions and water fluctuations probably affect fishing in Lake Tarpon more than the change of seasons does, there are definite patterns of location and food/bait preferences that fish follow throughout the year. Bass are particularly sensitive to seasonal changes, and they tend to act differently during the three major seasons on Lake Tarpon: Winter (December – March), Spring/Summer (April – September), and Fall (October-November). Transitional periods between these seasons vary yearly, as a function of temperature and rain conditions. We recommend checking with local guides and fishermen to determine which patterns the fish are "on", if you are fishing during season-to-season changes.

In the following sections we'll provide maps to show the hot spots for bass and crappie during each of the three lake seasons. *(Use the map in Chapter One for additional depth and location name information.)* Anglers who are fishing other species, particularly bream and catfish, won't need seasonal maps; these fish tend to stay in the same locations year round. The sections on each of these fish in Chapter Two, "Lake Tarpon Fish", provide information about favorite habitats and feeding habits for these and other species.

We'll also suggest specific baits and lures, and provide a few tips you can use to catch more fish in each different season. Some of these pointers have been offered previously in "Capt. Lenny's Fishing Report", published monthly in the *St. Petersburg Times.*

Winter (*December – March*)

Bass

Refer to the Winter Hot Spot Map for five excellent locations for catching winter bass in Lake Tarpon. These are in Anderson Park, Little Dolly Bay, Dolly Bay, Piney Point and Doc Red's.

These spots have similar characteristics. They all contain heavy vegetation (mostly hydrilla) and are in three to six feet of water. Most are near drop offs or sand bars where fish can move back and forth between deep and shallow water.

Shiners are most widely used to catch winter bass, though plastic worms, spinner baits and crankbaits are also successful. The premium shiners to catch trophy bass are Golden Lake Shiners, ranging from 6-10 inches in length. The bigger the shiner, the fewer bites, but (usually) the larger fish. If you want to catch more (but, on average, smaller) fish, use "commercial" shiners (called "pit shiners" locally). These shiners range from two to four inches. Occasionally they will attract a big bass, but usually entice fish weighing less than four pounds. Combining both commercial and lake shiners when fishing multiple rigs can give you the best of both worlds: the chance to catch a large number of bass, as well as the opportunity for potential trophies

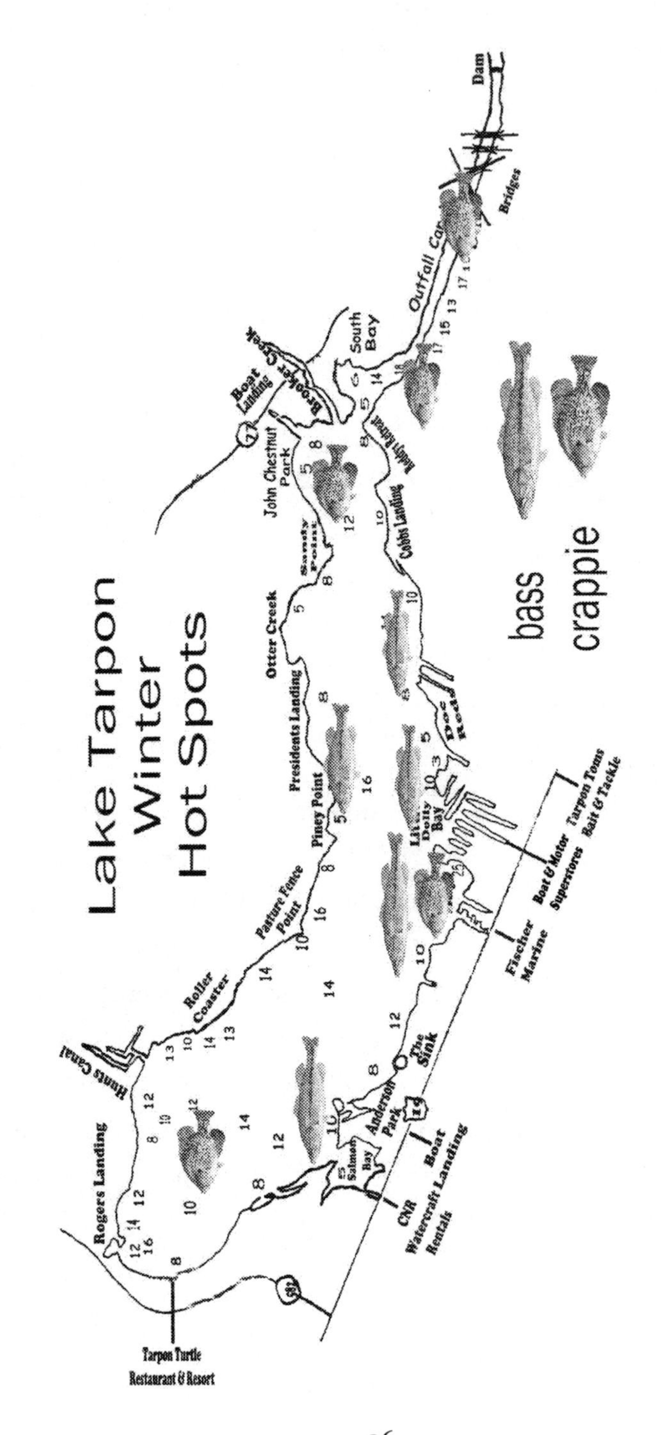
Lake Tarpon
Winter
Hot Spots
Rogers Landing
Hunts Canal
Roller Coaster
Pasture Fence Point
Piney Point
Presidents Landing
Otter Creek
Sandy Point
John Chestnut Park
Boat Landing
Brooker Creek
South Bay
Outfall Ca
Bridges
Dam
Reddys Retreat
Cobbs Landing
Doc Reds
Lil Dolly Bay
The Sink
Anderson Park
Salmon Bay
CNR Watercraft Landing Rentals
Boat Landing
Fischer Marine
Boat & Motor Superstores
Tarpon Toms Bait & Tackle
Tarpon Turtle Restaurant & Resort
bass
crappie

Crappie (Speckled Perch)

The Winter Hot Spot Map also shows five top areas to catch speckled perch in winter. These are all in some of the deepest holes and channels in the lake. Winter is the best time to catch lots of "specks" , some approaching two pounds! Anglers fishing Lake Tarpon for specks for the first time may be surprised that this variety of fish isn't found exclusively near structure or sunken trees. Fish finders and depth finders can help locate spots where there are heavy concentrations of specks, once you are in the general area where you plan to fish. The key to catching lots of specks is to find not only where they are located, but the depth in which they are suspended. Most of the time specks tend to stay tight in schools, often suspended literally on top of each other. Once you find the place and depth, the rest is relatively easy.

Missouri minnows (about one inch long) are the best bait for crappie year round, but they are especially effective in winter months. Small spinning lures (e.g., beetle-spin) also are effective, especially when you find fish suspended in open water areas like Dolly Bay.

Cane poles are the favorite tackle for most Lake Tarpon speck anglers. They usually set several poles in a semi-circle around the boat, after it has been anchored some 10-20 feet from the "speck hole." (Make sure you position the boat so that the drift will be toward the hole. Nothing is more frustrating than anchoring over a hole, only to find that the wind is carrying the baits away from the target area.) Spinning rods work well, too, with either live bait or artificials, and allow you to anchor the boat farther from the hole. This helps you avoid "spooking" the fish.

Spring/Summer (*April – September*)

Bass

The Spring/Summer Hot-Spot Map shows five excellent locations to find lunker (larger) bass during this season. These are in South Bay, Anderson Park, Sandy Point, Pasture Fence Point and the Outfall Canal. During the warm months of spring and summer, the majority of bass are caught early in the morning and just before dark. Though there are greater numbers caught during these periods, most lunker bass are caught in deep water during mid-day.

When fishing mornings and evenings, look for baitfish on the top of the water, and for "schooling" bass. Groups of smaller bass (from ½ to 3 lbs) often chase bait just as the sun comes up or just before dark. They offer anglers great action and are relatively easy to catch. When you see the schools, cast just beyond them using either a top water lure, small shad jig or small silver spoon with a trailing treble hook. If one of these lures doesn't get a strike, keep changing until you find the lure that most closely resembles the bait the fish are chasing.

At times, the size of the bait is critical. This is especially true if the baitfish are small. A lure larger than the bait will not get many strikes. The rule is to go smaller first, then work up, depending on strike ratio and the size of fish in the school. Once you find the right lure, you can often catch 20-30 fish in an hour. At times we've caught over one hundred bass a day, when they are schooling!

Lunker bass provide more of a challenge in the warm months. Though they can be caught, most of them are found in deepwater holes, often near structure. The top five spots identified for spring/summer all contain holes where big bass tend to lurk. The challenge is finding their feeding periods, and discovering the bait/lure they prefer. The plastic worm, rigged Florida or Texas Style, is the most frequently used lure to fish the deep holes in summer. (Florida rigs use a screw-in sinker, which keeps the sinker from moving; Texas rigs use a loose sinker, that moves up and down a little during the retrieve).

Worms come in literally hundreds of colors, styles and sizes. Most now have a built in scent – so spraying fish attractants is not necessary. Many fishermen fish worms too fast, because of experience using other lures. However, the majority of big fish strike either when the worm is lying still, or just after it has been pulled through brush or weeds. The ideal way to fish worms is to make the cast and let the worm sit still for a few seconds. Reel up all the slack, and slowly lift the rod perpendicular to the water (high overhead). Then lower the rod tip, slowly lift the rod perpendicular again, taking up the slack slowly, so you can still "feel" the worm on the bottom. When you feel a "tap" quickly take up all slack, lower the rod and rare back as if you want to break the rod. Many times you'll come up empty, or with weeds or branches, but the other times you'll set the hook on a nice bass. At times, the bass will take the worm and swim sideways or away from the boat. Other times, they will head straight for the boat, and you will feel almost nothing. If you think that the bass may have taken the worm but you don't feel anything, keep reeling fast until you feel tension on the line. Don't give up until you either see the fish coming toward the boat, or see the lure surfacing.

Some anglers prefer crank baits - deep diving - to go the bottom and stay near the brush during a retrieve. These lures tend to get hung up more often than worms, but do catch large numbers of nice fish. The retrieve of the crank bait is important, both in speed and style. Usually a medium retrieve with no jigging or jerking is preferred.

Crappie (Speckled Perch)

Crappie fishermen often skip the warm months and go for other species. However, crappie can be caught during the warm months – usually in deep water in the middle of the lake. Five productive areas for warm weather crappie are found on the Spring/Summer Hot Spot Map. The key is to drift live minnows slowly across deeper areas, varying the depth of the minnow, using a floating bobber. Once you get a strike or catch a fish, throw out a marker, and circle around and anchor so that the drift will let your bait approach the marker.

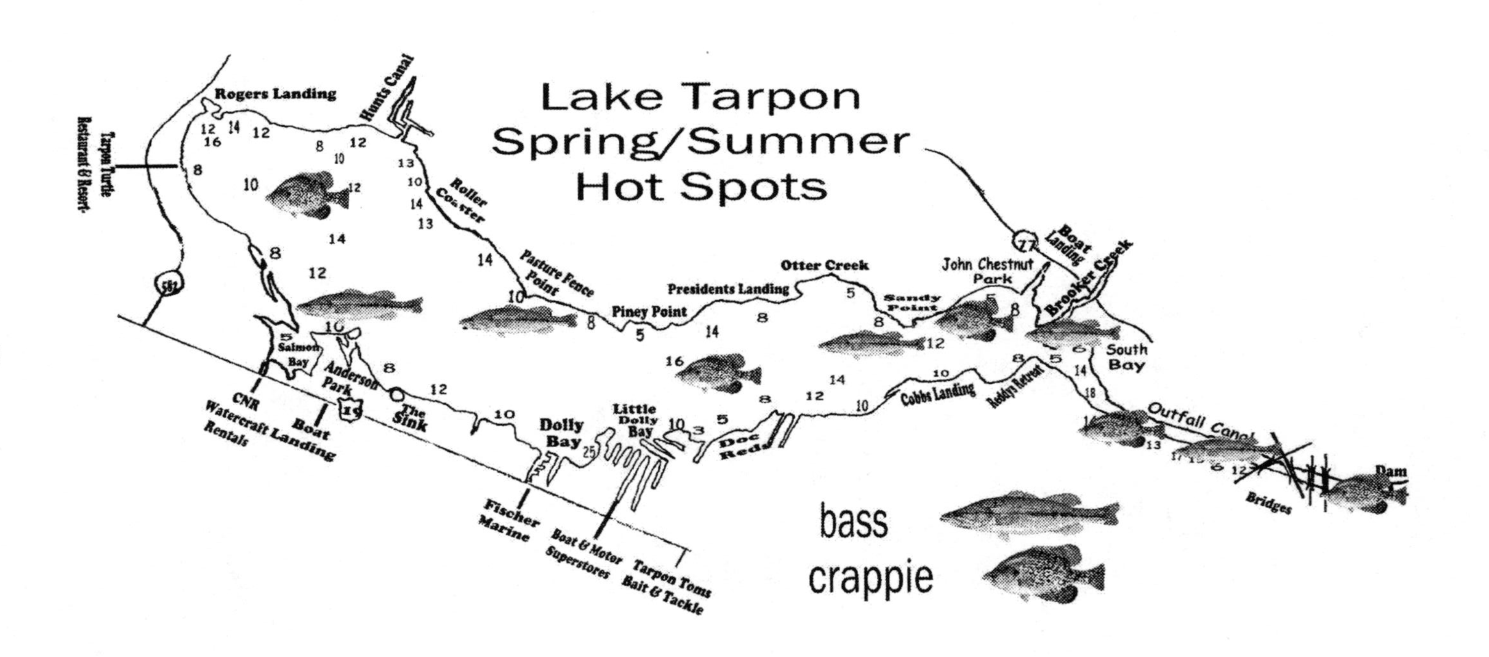

Lake Tarpon
Spring/Summer
Hot Spots
Tarpon Turtle
Restaurant & Resort
Rogers Landing
Hunts Canal
Roller Coaster
Pasture Fence Point
Piney Point
Presidents Landing
Otter Creek
Sandy Point
John Chestnut Park
Boat Landing
Brooker Creek
South Bay
Outfall Canal
Bridges
Dam
Salmon Bay
Anderson Park
The Sink
Dolly Bay
Little Dolly Bay
Doc Reds
Cobbs Landing
Reddys Retreat
CNR
Watercraft
Rentals
Boat
Landing
Fischer
Marine
Boat & Motor
Superstores
Tarpon Toms
Bait & Tackle
bass
crappie

Stay there until the bite stops, then repeat the process in other areas. If you use artificial crappie jigs, try slowing down the retrieve speed in summer, since the warm water makes fish move a little more slowly.

Fall *(October-November)*

Bass

In fall both shiners and artificial baits are productive, depending on water temperature and weather conditions. Overall, some of the top spots to fish for big bass include Anderson Park, Otter Creek, Dolly Bay, Little Dolly Bay and Reddy Retreat. These areas are pointed out on the Fall Hot Spots Map.

In warmer periods during the fall, artificials such as lipless crankbaits, spinnerbaits, and spoons are effective. Texas and Florida style rigged worms also are productive. During cold snaps, when the water temperature begins to go down, shiner fishing will tempt the lunker bass. Schooling bass also can be found in South Bay and the Outfall Canal early in the fall. Shad colored jigs with an 1/8 ounce lead head are especially productive with the fall "schoolies."

In many ways, fall is a transition period in the lake, and top anglers adapt their lures/baits and approaches to each day's individual weather conditions and water temperature. As fronts pass through, baits and approaches even need to be changed several times during the day!

Crappie (Speckled Perch)

As the temperatures drop in the fall, crappie fishermen are seen on the lake in greater numbers. Limit catches of crappie are caught frequently in late fall on Lake Tarpon. The Fall Hot Spots Map shows some of the best places to look for specks. Once again, deep holes are the favorite location, particularly those near some structure. Bait, tackle and techniques are the same year round for crappie. Once you locate where they are stacked, you can count on catching quite a few, given proper weather and water temperature.

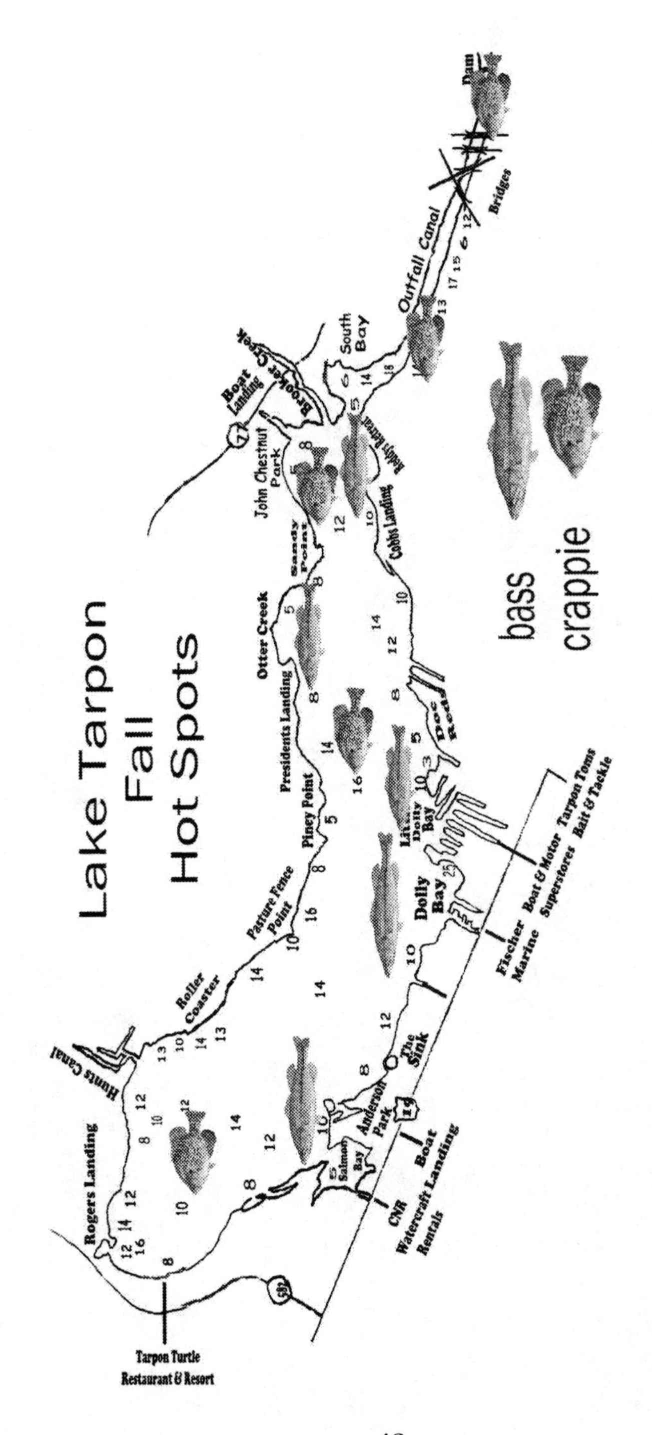

Lake Tarpon
Fall
Hot Spots
bass
crappie
Dam
Bridges
Outfall Canal
South Bay
Brooker Creek
Boat Landing
John Chestnut Park
Sandy Point
Cobbs Landing
Otter Creek
Presidents Landing
Doc Reed
Piney Point
Dolly Bay
Boat & Motor Superstores
Tarpon Toms Bait & Tackle
Fischer Marine
Pasture Fence Point
Roller Coaster
Hunts Canal
Rogers Landing
The Sink
Anderson Park
Salmon Bay
Boat Landing
CNR Watercraft Landing Rentals
Tarpon Turtle Restaurant & Resort

Chapter Four

Fishing Tips and Techniques

Chapter Four

Fishing Tips and Techniques

Introduction

There are literally millions of books, articles and web pages available that offer excellent fishing tips and techniques. We know that the readers of this book will be familiar with many of them. Therefore, our purpose in this chapter is simply to focus on a few areas that uniquely apply to fishing in Lake Tarpon. The tips we have included in this chapter relate to shiner fishing, spinner bait fishing, crankbait fishing, plastic worm fishing, and crappie fishing. We want to thank our contributing editor, Dan Richardson, for sharing some of his field tested fishing secrets about using spinner baits and crankbaits.

Tips on Fishing Plastic Worms

Every fisherman has a bait he relies on when conditions are tough. We refer to this bait as a confidence bait. Our confidence bait is a plastic worm. Over the years we have caught more bass on a plastic worm than on any other bait in the box.

One of the most important choices you will make is the rod you select to fish plastic worms. The rod needs to have backbone, so you both get a good hook set and have sufficient sensitivity to feel the bottom cover and the "bite". The reel should be capable of holding 12 to 15 pound test line. . . and a smooth drag is a must.

Most of the year bass are found holding in isolated cover – either grass or wood. Therefore, we throw a Florida rigged plastic worm most of the time. When we rig the plastic worm we take into consideration the water depth, wind , current, water clarity, water temperature and available cover. These factors will determine line size, the size of the worm weight and the color and style of worm to choose.

Beginning with line choice, if the water is dirty, with bottom cover to snag, 15 pound test line is best. In clear water we will drop down to 12 pound test line. Weight selection is also important. One rule of thumb is if you can't feel the worm, you will not feel the bite! Therefore, use enough weight to clearly feel the bottom. In shallow water, with very little wind, a 1/16 ounce or 1/8 ounce weight will often be our choice. However, when fishing in deeper water, or in windy conditions, a 3/16 or ¼ ounce weight is generally best.

In considering hook size, we take into account the size worm we are using. With a 4 inch worm, a 1/0 seems to work the best, while with a 7 inch worm a 3/0 to 4/0 fits better. With worms 10 inches or longer, a 5/0 would be our choice. A good rule here is: if it doesn't look like the hook fits, don't use it. Often a hook that is too big will impede the worm's action – and a hook too small will not allow you to get a proper hook set.

When choosing the style of worm to fish (e.g., straight tail, flat tail, curly tail, etc.), water temperature is the number one consideration. Normally, in cold water, bait fish and bass are less active. Therefore, under these conditions, we use a straight tail worm that gives off little vibration. When the water temperature warms up and bait fish become more active, we try to match this activity by using a worm that gives off more vibration. A ribbon tail worm works best in this case.

In selecting the color of the worm, take water clarity into consideration. In clean water, clear or translucent colors tend to look more natural. In dark or stained water, dark colored worms offer a better silhouette for the bass to see. In general, natural colors tend to work in both types of water.

We take into consideration the season and migration patterns when choosing the length of the worm. During the early part of the year, bass are nesting and often will grab the tail of the worm to move it off a nesting area. At these times, we prefer a smaller worm, which allows the fish to get the business end of the bait closer to its mouth.

As the year progresses, bait fish get bigger and we want to match the size of what the bass are feeding on. Our number one choice in warmer weather is a 7 – 8 inch worm. While big worms tend to get big bites, and small worms tend to get small bites (nesting fish being the exception), through the years we've found that a 7 - 8 inch worm catches a better average size fish than the shorter or longer baits.

We use several different rigs when worm fishing. These include Florida rigs, Texas Rigs and Carolina Rigs. With the Florida rig, the worm weight is screwed into the head of a plastic worm, which is rigged weedless on the hook. With the Florida rig, you cast the bait toward the type of grass or cover the bass are holding in that time of the year. Then, allow the bait to free fall to the bottom while holding your rod tip high. If you see a hop in the line or feel a tap, reel down till you feel the weight of the fish and then set the hook – hard! In this case, the bass just took the bait on the fall.

If nothing happens when the bait is falling, allow it to reach the bottom, keeping your rod tip in the 10 o'clock position. Then use your reel to take the slack out of your line. We never allow the rod to go lower than the 10 o'clock position unless we're setting the hook. To move the bait, just lift the rod tip 2-3 inches. This will allow the worm to slide across the bottom. Often, novice fisherman want to set the hook when they feel bottom, because they think that they are getting a bite. If you are moving the bait, you're probably feeling the bottom – not a fish. Look for the bite when the bait is resting. When you get to the 11 o'clock position, lower your rod back down to the 10, take up the slack with your reel so that you maintain a tight line with the worm, and feel the bottom all the time. When you get a bite or tap, lower your rod so it's parallel to the water, reel up your slack, and set the hook hard. A good rule is when you feel a tap, reel down as fast as you can and set the hook. Some anglers believe you need to allow the fish to swim off with the bait before trying to set the hook. We feel that, since the bait is not real, the longer the bass has it in its mouth, the better chance it has to realize something's wrong and spit it out.

Another effective method of worm fishing uses the Carolina rig. Our favorite time to use this type of rig is during the migration periods (Spring and Fall). During these periods, fish are roaming much of the time. They cruise around points, bars, and channels. The Carolina rig allows you to keep the bait in the strike zone for longer periods of time.

To rig a bait Carolina style, start with a rattling Carolina rig sinker. In selecting weight, consider the depth of water you will be fishing and the current. A ½ ounce is a good starting point for sinker selection. Place the sinker on the line and slip a bead below it, then tie a barrel swivel below them both. To the barrel swivel tie a 12-18 inch 12 pound leader (we prefer fluorocarbon for the leader, since it is less visible to the fish.) Then tie an offset wide gap hook- - 1/0 to 4/0 to the end of the line, depending on the size worm you'll be casting. Finally, slip the worm on the hook. The reason we use an offset hook is that, when you cast the Carolina rig, the worm is going to flip like a whip. The offset shank will prevent the worm from sliding down the hook during the cast.

To work the Carolina rig, cast it onto the structure or bottom contour you suspect that the fish are holding in or over. Again, keep your rod tip high, and allow the bait to hit bottom before beginning the retrieve. Then slowly drag the worm with the rod and take up slack with the reel, pausing briefly between each drag of the rod.

When a bass bites the Carolina rig, often you will not get that "thumping" sensation you may be accustomed to when using Texas or Florida style rigs. (Texas rigs are identical to Florida rigs, with the exception that the worm sinker is not attached to the worm, and moves free along the line). When Carolina rig fishing, the bite will often be very soft and may feel like a "wet rag". If you put a little pressure on the line when you get that feeling, you may find life at the other end of the line. When this happens, set the hook!

When Carolina rig fishing, don't be surprised if you don't feel a bite, but you see a fish is swimming off with your bait!

Artificial scents allow a bass to hold onto a bait longer. So when we are missing some fish after feeling bites, we use artificial scent to tempt the fish to hold on longer.

Don't be discouraged if you have limited success when fishing these baits for the first or second time. Worm fishing is definitely an art – one that can be mastered only with practice. However, with repetition, you'll develop skills that will lead to your catching more and bigger bass.

Spinnerbaits

by Dan Richardson

Often overlooked but never forgotten, a spinnerbait is very productive on Lake Tarpon. The key is to properly fish the bait. A consistent tournament winner known as "Top Water Charley" uses only one lure, a twin propeller bait known as a Devil's Horse, when fishing for bass. One lure, one technique, one color. The reason he uses only this lure is that he is successful and consistently catches sufficient bass to finish in the money.

I have been called, "Spinnerbait Dan," not because I only use a spinnerbait, but because I have become proficient with this lure and consistently use it to catch bass and win tournaments. This section discusses Types, Bait Size, Color, Areas to Fish, Speed of Retrieve and Equipment. Read on, it will help you catch more bass on spinnerbaits!

Types of Spinnerbaits: There are essentially two types of spinnerbaits, in-line and safety pin. Both are effective when used under the proper conditions. In-line spinnerbaits are excellent in heavy cover such as thick reeds, fallen trees and hydrilla. They do not have a tendency to get hung up, because the hook follows the blade that protects it from obstructions. The safety pin type is good in lighter cover and more open areas, and is the type most often used on Lake Tarpon. The key to catching fish on this bait is to target reaction strikes. It is most effective in depths of less than five feet and areas that have some type of cover such as hydrilla, reeds, and stickups.

Bait Size: Matching the hatch, meaning using bait that most closely resembles the size of the forage most consumed by bass, is a proven concept. This holds true for spinnerbaits; however, matching the weight of the spinnerbait is not the only consideration. The blade is the key. I primarily use hammered nickel blades, both willow and Colorado shapes. I use willow in cleaner water where the added vibration is not needed. I use the Colorado blade in dingy or muddy water because it puts off more vibration. Most of the time I use a 3/8 oz bait and will move to smaller or larger baits if there is a lot of fishing pressure or if I am not catching fish.

Color: Spinnerbaits come in all the colors of the rainbow and in combination colors, too. I use, white, white, or white. This may seem comical, but it is true. I have been fishing with this lure for over 25 years and with few exceptions, white catches as many, if not more, than any other color I have seen used. I feel color is not what attracts the fish. It is spinning blades and vibration that get their attention and cause the strike. White can be seen in murky water and is not too offensive in clear water. It is universal and it works in any state.

About three years ago, I found I was catching fewer fish on spinnerbait. I changed retrieve speed, rod length, reel ratio and did not increase my catch. Then I pulled out an older bait and realized that the skirts on the new baits have changed. Although still coded as white, white skirts are being substituted with translucent or off colored skirts.

This is a critical change. I keep a supply of white rubber replacement skirts on hand, and put them on all my baits. I am back to consistently catching fish. I keep a spinner bait or two with the translucent skirt to use in very clear water, but they are not often used.

Areas to Fish - Cast Parallel to the Cover! This is the most critical element. Imagine fishing a reed line with your boat positioned about a cast length from target. Typically one casts the lure to the cover and steadily retrieves it to the boat. Unless the fish are holding off the cover, this method quickly moves the bait out of the strike zone. Depending on the clarity of the water, the strike zone can be anywhere from six inches to three feet.

I suggest casting parallel to the cover. This is done by positioning your boat a few feet from the cover, making long casts and retrieving your bait close to the cover. This method allows you to show your bait to more fish, thus increasing your chances of getting bit.

Speed of Retrieve is Critical: How many times have you heard stories about anglers who catch large numbers of bass in the early morning hours, but by about 10 a.m., have stopped catching fish? Many attribute this to the bite stopping. This is not always true. What typically happens is that the fish move to a little deeper water or down into the cover they came from to feed. Many anglers are aware of this and slow their retrieve allowing the bait to move lower in the water column. This is not good. I have found fish hone in on a certain speed and will hit at that speed throughout the day. In the early hours, I vary my retrieve speed about every ten casts until I find the magic retrieve. Then as the day progresses and the fish move either into deeper water or into the cover, I reduce either the number or the size of the blades. This can be done by clipping the front blade, or by changing baits. A smaller blade has less resistance; therefore, the bait will swim lower in the water.

Equipment: I use only two spinnerbait rods, a 5.9 ft medium action Marado rod and a Marado 7 ft medium action Terminus rod. On both rods I use the Marado Icon 600 real with a 5.1:1 retrieve. This 6ft rod is used in tight areas or when the presentation is critical. With a shorter rod, I am more accurate and can get the bait into the water with little to no splash. In open areas, I use the 7ft rod. I always use at least 15-lb. mono and on occasion, I use heavier line. The line size is not important, because you need not impart any action on the bait. I will caution you about using little or no stretch lines, I have tried them and my catch ration dropped. The reason is with mono, the bass gets more of the bait into its mouth because of the stretch. With superlines, this is not the case.

The Art of Fishing Crankbaits

By Dan Richardson

Picture your grandfather's old tackle box with the beige bottom and two green plastic lids. On either side of the lids were latches and below were two pull out trays that opened like staircases on opposite sides. There was cork on the bottom of each compartment where the different surface and sub-surface lures were neatly displayed. Then recall what your grandfather used most often; I suspect it was either the surface lures or the rubber worms. Perhaps, like my grandfather, yours would cast a sub-surface lure a few times and then return to his confidence bait, which was a surface lure. On Lake Tarpon, or any bass-ridden lake, the crankbait is a critical part of an angler's arsenal. There are two types of crankbait: lipless and lipped. I will discuss both.

Lipless Baits

Lipless Baits. This includes the Bill Lewis Ratt-L-Trap, the Cordell Spot and the L&S (Rock'n Rattl'r.) These are used for depths typically less than five feet and work best around eel grass, hydrilla, or other aquatic vegetation. The lure is worked by making long targeted casts to openings in the cover and then rapidly retrieving the bait to the boat. The strike can come at any time, but most often occurs early or mid-way through the retrieve.

Speed of Retrieve. On any given day a bass will strike a lipless lure at different speeds; sometimes a fast retrieve and at other times, a slow retrieve. Vary your retrieves throughout the day until you find the most effective one. Remember that most strikes come after contact with vegetation; therefore, touching the vegetation is critical to success. If you are not contacting the vegetation, you are not getting most of the strikes. I like to cast into the vegetation, allow the bait to partially hang, and then rip the lure free by sweeping the rod to either side. This causes a reaction strike and a fish will strike the lure even if it is not hungry.

Color. In clear lakes, chrome finished lures with either a black or blue back are most popular. Do not get worried if the paint chips off the bait; many pros feel, as do I, that an older lure works better because it resembles an already injured baitfish. In dark or muddy water, firetiger, fluorescent yellow with a blue back or something with a red hue are colors of choice. Do not be afraid to experiment with different colors. I recall a tournament I fished on Lake Tarpon a few years ago. Many of the tournament anglers were successfully catching keeper bass using chrome with a blue back. I switched to a new color sent to me by L&S Baits, maker of the Rock'n Rattl'r. The color was RT, which stands for rainbow trout. Now, I don't know about you, but I grew up on Lake Tarpon and have fished it for the past thirty years. I have never seen a trout on the lake. Why on earth would a bass bite such a color? Simple: it's different then what they are seeing. I went on to catch a limit of bass, had two kicker fish and won the tournament. While I am convinced I could have caught a limit of bass using the traditional chrome and blue, because I switched I was able to catch better quality bass.

Bass become bait shy. Studies show bass have the ability to recognize lures. They become less attracted to a lure they see day in and day out. This is why I keep several different colored lures in my box. I often practice with one color and compete with another. A friend asked me the other day, " What is your favorite color lure?" I responded, "In practice, chrome with a blue back." "What about on tournament day?" he asked. "I use several colors," I responded. I will, of course, start with the lure that caught fish in practice, but I will quickly change if the bite slows or if I want better quality fish.

Size. Lure size is one of the most important choices an angler makes. It contributes to the number and the quality of the creel. "Match the hatch" always comes to mind, that is, find a lure that most closely emulates the forage. Lipless crankbait vary in size including ¼, ½, and ¾. If a lake is receiving heavy fishing pressure, I suggest using either a ¼ or 1/2 oz lure. If you have the limit of bass and you want to upgrade you catch, increase the lure size.

Lipped Baits

Lipped crankbaits are either super shallow (less than 1 foot), shallow (1 to 3 feet), medium running, (3 to 6 feet) or deep running (greater than 6 feet.)

Speed: Speed is critical to maintain a certain depth, but not as important as it is with lipless baits. The crucial part is finding the depth where the bass are feeding. Let's say you find fish on your recorder holding at 6 feet. Logically, one would use a lure that runs 6 feet. Be careful! The true depth of a crankbait is determined by speed of retrieve, line diameter and the position of the rod. A lure that is said to run 6 ft could run anywhere from 6 feet or less. If you err, err on the shallow side. For example, if bass are holding at 6 feet, they usually feed at that depth or less. If one throws a lure running deeper than 6 feet, the fish will not see the bait. Remember the swing effect, too. Unlike lipless baits, which sink and can be counted down to certain depths, crankbaits must be retrieved to the desired depth; therefore, they are only at the maximum depth for a short period of time. Based on this, it is critical to cast well beyond the targeted fish to allow the bait to reach optimum depth before it passes the bass. Holding the rod in a low or down position allows the bait to stay deeper longer. Conversely, if one wants the bait to run shallower, one should raise the rod, causing the bait to run higher in the water.

Size: This separates the fishing from the catching. I have reviewed many books, articles and videos, and I am convinced the size of the lure is absolutely critical in catching large numbers of quality bass. On Lake Tarpon, for example, a lower profile lure consistently does better than larger baits. The reason is that the shad are typically less then 3 inches in length and closer to 2.5 inches. When throwing bait, one must show the fish the lure size that matches what it is eating. My most productive bait is the L&S Shad Rattler 91LSR, color number 21. The reason it is so good is that it most closely matches the typical shad in lakes around Florida. Based on this, I am not suggesting that you open your tackle box and eliminate baits larger than three inches. They have their place.

If you are fishing deep murky water, larger lures are more easily seen and, since they often put off more vibration, are easier to find in the water. Large baits are also good for catching large bass. So when you get a limit and want to upgrade your catch, use larger bait.

Color: The factors are essentially the same as those with lipless baits; I use only a few different colors: firetiger, shad colors, and something with red. Based on my personal experience, color is not as critical as size and keeping the bait at the proper depth.

Relationship to Cover: This is the most critical factor when fishing crankbait; the bass must be related to the cover. Many times one will travel around the lake and see suspending bass on the recorder. This is typical in Lake Tarpon in areas such as the South Canal, North Cove and the various structure piles around the lake. If the bass are suspended, (holding between the cover and the surface), you are probably not going to be successful in catching those fish. Suspended bass are the most difficult to catch, and most pros do not actively fish for them. The reason is simple: suspended bass are not feeding. They are suspending, which means they are not hiding to ambush prey. Look for bass relating to cover and you will find bass that can be caught.

Equipment: Getting the strike is only half the battle; landing the bass is another story. I suggest using a rod specifically designed for cranking. I use a 7 ft long fiberglass rod with a soft tip and plenty of backbone. While it may be heavier than graphite, the benefits are many and I do not lose as many fish during the fight. A good quality baitcasting reel with a 4.7 or 5.2 retrieve gets the job done.

There is a lot more to fishing crankbaits than just throwing at toward the structure. Here is one approach:

I frequently fish the Pipeline on Lake Tarpon. The Pipeline is located on the northeast side of the lake just about ¾ mile north of Salmon Bay (Anderson Park). The Pipeline runs from the bank to about 11 feet of water, and is a typical water pipe made of cement. It is about 5 feet in diameter and in the shallow water is under the sand. Its deepest point is about 3 feet above the bottom.

Many times on my recorder, I will see bass holding above the pipe and when this is the case, I am not very successful. When the bass are holding on either side of the pipe, such as north or south, I typically catch a few. If the fish are holding below the pile, watch out! - I can load the boat with quality bass. This is because the fish are relating to the cover on either side of the pipe, and are willing feeders. When they are below the pipe, they are in the ambush position and are ready to attack.

Here is how I would fish the cover. First, I would pass over the pipe and mark it in about 7 feet of water. The reason I mark it at 7 feet is because the pipe is most easily found at this depth, and I can fish either the shallow side or the deep side without having to fight my marker. The pipe ends at about 11 feet, depending on the water level. Years ago the end of the pipe was marked with four telephone size poles. There was a water level indicator on one of the poles.

.After a few near fatal boating accidents, the poles were cut below the water surface. These poles are referred to as the platform, and many anglers fish only this section of the pipe. If you find the platform, you will probably catch fish.

The problem with fishing the platform is that fishing and anchor lines are entangled around the poles and if contacted, a lost lure is not uncommon. At last count, I have lost 56 lures on the pipe. At $5 to $7 per lure, the loss adds up and it is disappointing. Now with many lures costing north of $15, it could even be depressing!

I recommend carrying a lure retriever; it will pay for itself many times over. I start by positioning my boat about fifteen feet to the south of the pipe and cast as far north as possible. This will allow the bait to get to the proper depth before it contacts the pipe. I typically am successful with one of three retrieval methods. First is a steady retrieve; once I have contacted the pipe, I kill the bait, allowing it to slowly rise to the surface. The second method is a steady retrieve, letting the bait deflect up and over the pipe. This gets the reaction strike. The last method is slowly dragging the bait over the pipe and once it is off, speeding up the retrieve to emulate a fleeing baitfish. This method consistently works. After fishing to the north I parallel cast on either side, always making contact with the pipe. I also turn off all electronics. The signal from either a flasher or recorder is a clicking sound and this bounces off the pipe.

Remember: if you catch a bass, do not be surprised if you do not immediately catch another. Keep throwing and give the bass a chance to regroup. Also try throwing the same lure in a different color, or changing the lure size. If you master lipless and lipped crankbaits you can consistently catch bass all over the United States!

Shiner Fishing Tips

A major key to shiner fishing success lies in keeping the bait alive and healthy.

Make sure Lake Shiners you buy from local bait shops have been "cured". Curing is a process involving multiple stages over several days to help the shiners stay alive, active and aggressive longer after you have bought them. Shiner suppliers usually sell only shiners that have already gone through the curing process to bait shops. Ask the clerk at the bait shop if their shiners have been cured.

You can also catch shiners in the lake, using a cane pile, a doughball (mixed with cotton to help it stay on the hook) and a number 10 hook. Most shiner fishermen "chum" a spot with dry hog feed about 30 minutes before fishing for shiners. Ask local guides or fishermen about some of the best places to catch shiners in the lake. Cast nets also can be used to catch multiple shiners from "chummed" holes. However, keep in mind that shiners you catch in the lake are not "cured" – and may not last more than one day, even with aeration.

Once you have your shiners, always use aerated containers to keep them alive and active. Adding a few ice cubes or a scoop of crushed ice can also help add oxygen to the shiner water. You also may want to add some commercial preparation to help condition your water – similar to that used by tournament fishermen to keep bass alive in their live wells during the day of the tournament. Several different brands of water conditioners for keeping fish alive longer are available in local tackle shops.

Shiners will jump out of your container any time at every opportunity – so always keep it covered. Remember to keep dip nets covered with your hand, if you are transferring shiners from one container to another. If shiners jump out of the container and land on a hard surface, they often injure themselves and die quickly.

When rigging shiners that will be drifted or trolled, it is usually best to bring the hook up from the underside of the shiner's mouth and let the point come through one of the shiner's nostrils. This rig allows the shiner to swim naturally and doesn't stress it unduly. When planning to anchor and let the shiners swim around a specific area, some anglers prefer to hook the shiner through the back, just under the top (dorsal) fin. This rig tends to let the shiners swim more naturally when being drifted or trolled.

Remember to change hook sizes according to the size shiner you are using. In no case, however, should you use a hook smaller than 3/0. 5/0 hooks are the standard for rigging the large lake shiners.

Most shiner fishermen use corks or bobbers when drifting or anchoring up on holes to fish for bass. We've found that drifting or trolling a "live line" (also called "free line") can be effective at times when bass won't take shiners on a bobber. Free/live lining involves letting the shiner swim freely without a cork, and keeping some tension on the line to feel the bite. When a bite is felt on a free line, the spool is left open (with clicker off) for a few seconds to let the bass run and get a good hold on the shiner. Then the hook is set. There are problems with underwater vegetation and structure, which can occasionally cause the freelined shiner to get hung up, but that's where the bass are hiding too . . .so it's worth the extra effort.

Partially inflated balloons as floats are also used in shiner fishing. They work best when fishing around structure, lily pads and reeds. If the bass runs into the cover the balloon pops and helps increase the chance of landing the fish.

Crappie Fishing Tips

More crappies are caught on minnows than any other bait. Keeping minnows alive is not hard in winter, but remember to use an aerator in the bait bucket if it is kept in the boat. Otherwise, use the flow-through minnow buckets that hang on the side of the boat and float on the surface. If you have a bucket with no aerator, putting ice cubes or a scoop of crushed ice in the minnow bucket will help add oxygen for a short time. Pellets that add oxygen to help keep Missouri minnows in buckets alive are also sold in bait shops. (Missouri minnows are the most common crappie bait – averaging 1 to 2 inches in length).

It's usually best to use a slip bobber when fishing Missouri minnows. The bobber is allowed to float at the level set by a stopper (usually a knot tied on a piece of string attached to the line). The slip bobbers allow anglers to cast the minnow long distances with ease. The rest of the rig is a number 4 offset hook (gold ones seem to attract more specks) and a small splitshot sinker placed about 12 inches up from the hook.

Cane poles are the favorite tackle for most Lake Tarpon crappie anglers. Locals usually set several poles in a semi-circle around the boat, after it has been anchored some 10-20 feet from the "speck hole". If you anchor near a prime spot, make sure you position the boat so that the drift will be toward the hole. Nothing is more frustrating than getting anchored up on a hole, only to find that the wind is carrying the baits away from the target area.

Spinning rods can be effective, too, using either live or artificial bait. Casting allows you to anchor the boat farther from the hole, thus helping you avoid "spooking" the fish.

Specks do tend to be wary at times, so using the lightest line practical is the best way to go. However, since you may catch an occasional bass, catfish, bowfin, or other lake dweller while crappie fishing, it's usually best to go with at least 6 lb test monofilament. With ultralight rigs, 4 or even 2 lb test monofilament is okay. Be sure to set the drag properly, Most fish lost on light line are due to improper drag set, rather than light line test.

One final reminder: if you use suntan lotion or insect repellent on your hands, don't forget to use a dip net to get minnows from the bucket. Otherwise, if you dip your hand in the bucket, the oil from the lotion will contaminate the water and you'll lose some bait.

Chapter Five

Fishing Records and Awards

Chapter Five

Fishing Records and Awards

Lake Tarpon has been cited in the past as the lake producing the second largest bass unofficially caught in the state of Florida. The Bassmaster Top 25 (*Bassmaster Magazine*, January, 2002, p. 54) cites a 19 pound bass caught by Riley Witt on June 21, 1961. (The largest Florida bass (20 lb 2 oz) was caught by Fritz Friebel on Big Fish Lake in May, 1923).

Captain Gene Zamba, a Lake Tarpon guide, holds a world record for largemouth bass on a fly rod at 13.5 lbs. This record bass was caught in Lake Tarpon in 1996.

Larry Barker, one of this book's authors, holds the all tackle catch and release world record for largemouth bass. His 26 inch Lake Tarpon bass was caught and released in 2002.

Given that the top weight in the *Bassmaster* Top 25 is 22.25 lb (caught on Montgomery Lake, GA, in June, 1932), it seems probable that another world record bass is now living in the depths of Lake Tarpon as this guide is being written! Keeping in mind the number of different line classes and fishing methods (e.g., fly rod, spinning, etc.) for which official records are held by the International Game Fishing Association, the National Freshwater Hall of Fame, the Florida Game and Freshwater Fish Commission and other organizations, it is possible that several current records are waiting to be broken on any given day by Lake Tarpon anglers!

Lake Tarpon records had not been kept officially until 2002. As of 2002, **Tarpon Tom's Bait and Tackle** began certifying and recording Lake Tarpon fish records. Different monthly and yearly records are recorded for largemouth bass caught on artificial lures and shiners. Annual records are also kept for specks (crappies), bluegill (including sunfish and shellcrackers), catfish, gar and bowfin (mudfish). Certificates are given for annual awards and new Lake Tarpon records. At present, no separate records are established for those caught on fly or spinning tackle, or for different line classes.

Lake Tarpon Records

The procedure for reporting and recording an eligible fish caught in Lake Tarpon is as follows:

- Bring the fish (alive if at all possible, so it can be released after the record is certified) to Tarpon Tom's Bait and Tackle during business hours.
- Complete the registration form (see Appendix A). A photo is required to be filed with the registration form.
- Release all fish (required for bass), ideally in the same area where they were caught, as quickly as possible after certifying the record.

Florida Fish and Wildlife Conservation Commission

Certified State Records

The process to apply for a Florida fishing record is as follows:

Certified state record fish must be legally caught using an active hook-and-line method (including a proper license or exemption), by sport-fishing methods, identified by a Florida Fish and Wildlife Conservation Commission (FWCC) biologist and weighed on a certified scale. To set a new record, you need to exceed the certified records at the time. If you believe that you have caught a record fish, the FWCC asks that you keep it alive or place it in ice water. They suggest that you do not freeze the fish because it will lose weight when thawed. As soon as possible, call the nearest FWCC office to arrange to have a fisheries biologist meet with you (addresses are available on their website: floridafisheries.com).

Big Catch Program

The "Big Catch" program gives recognition to anglers who catch fish that exceed the minimum qualifying weights and lengths established by the Commission. A complete list of minimum weights can be found on the Commission's website: floridafisheries.com/record. Anglers who catch a trophy-sized fish receive a color citation showing the type of fish they caught plus a window sticker. In addition, anglers who release their fish get special recognition from the program. An angler catching five "Big" fish of the same species will be recognized as a "Specialist" , five trophy fish of five different species qualifies as a "Master Angler", and five big fish of ten different species qualifies as an "Elite Angler."

To qualify for the award the catch must witnessed and a *Big Catch* form must be completed. Forms may be downloaded from the web site: floridafisheries.com.

At the time of this writing, the minimum weights in the adult class to qualify for a *Big Catch* award were:

Species	Total Length*	Total Weight
Largemouth Bass	24"	8.00 lbs
Black Crappie	14"	2.00 lbs
Bluegill	11"	1.25 lbs
Brown Bullhead Catfish	16"	2.00 lbs
Florida, Spotted Gar	28"	5.00 lbs
Bowfin	30"	10.00 lbs

**To determine the length of the fish, measure from the tip of the snout back to the tail with the fins squeezed together and the mouth closed*

Check the website address above to obtain a complete list of all species and qualifying lengths/weights for the *Big Catch* award. The site also gives a similar list for young anglers under 16, with slightly lower required lengths and weights.

Sample Big Catch Specialist Certificate

Florida Fish and Wildlife Conservation Commission
Angler
BIG CATCH SPECIALIST CITATION
presented to Larry Barker
For Catching Five Qualifying
LARGEMOUTH BASS
Division of Fisheries Director
Executive Director

International Game Fish Association (IGFA) Certified State and World Records

The International Game Fish Association is arguably the most prestigious and broad based of the agencies certifying fish records. Membership in the IGFA is not required to qualify for or hold a world record. However, anglers striving to break state or world records probably would want to join the IGFA in order to keep up with current rules, regulations, records, pending records, etc. The address of the IGFA is: 300 Gulf Stream Way, Dania Beach, FL 33004. (Phone: (954) 927-2628). You can become a member online by going to the website: igfa.org/memberships.

The IGFA has kept world records for many years, but only recently has been keeping individual state records. Freshwater records (both world and state) are recorded and certified for 4 lb, 12 lb and 20 lb line tests. (Note: the line test on manufacturers' labels does not always meet IGFA standards). The IGFA also certifies records separately for fish caught on fly rods, with several different tippet classes (see IGFA web site for relevant tippet classes).

Since the state records are relatively new for the IGFA, there are lots of them waiting to be broken. The 2001 IGFA records for bass in Florida all were in the 8 pound range. These could be broken any given day on Lake Tarpon. The same is true of other species such as bluegill, specks and catfish. All current records are clearly within range to be broken in the waters of Lake Tarpon.

A main reason that many record fish are not certified is that there is a lot of "red tape" involved in setting a record. (You can download a record application form from IGFA from their web site: igfa.org). However, if you are willing to read the rules and regulations in advance, and be prepared to meet the standards, record setting can be exciting.

In addition, the IGFA offers special recognition and certificates for all largemouth bass caught weighing over ten pounds. There are requirements to qualify that also may be downloaded from the IGFA web site. The award is called "10 pound bass club." Line samples, photographs, and witnessed weighing on certified scales are part of the requirements for all IGFA records and awards.

National Fresh Water Fishing Hall of Fame (NFWFHF) World Records

The National Fresh Water Fishing Hall of Fame and Museum in Hayward, Wisconsin, has been keeping records for a variety of freshwater species (including released fish) for almost 4 decades. If you want to join or find out about their current records, you can check their web site: freshwater-fishing.org. If you prefer to write, their address is: NFWFHF, P.O. Box 690, Hayward, Wisconsin 54843. The NFWFHF certifies records for a wide range of line classes and tackle categories, so there are many NFWFHF records ready to be broken by Lake Tarpon fishermen.

Sample World Record Catch and Release Certificate

OFFICIAL RECOGNITION CERTIFICATE

WORLD RECORD
CATCH AND RELEASE
Angling Achievement

National Fresh Water
Fishing Hall of Fame®

ATTEST BY THE GREAT SEAL

Larry L. Barker
ANGLER

BASS/Largemouth
SPECIES

3/22/02
DATE CAUGHT

26 Inches
LENGTH IN INCHES

Division #1 - Rod/Reel
METHOD OF CATCH

Lake Tarpon / Florida, USA
WHERE CAUGHT

EXECUTIVE OFFICER

Unlimited
LINE TEST CLASS

Other awards for trophy fish

In addition to the organizations noted above who certify records and offer recognition for trophy fish, numerous manufacturers offer certificates and/or patches to fishermen who catch trophy fish using their products. For example, Shimano, one of the more established tackle manufactures, offers awards for trophy fish caught using Shimano tackle. If you are interested in pursuing awards, check the Internet or write to see if the manufacturers of the tackle, lure or line that you used to catch the trophy fish offer similar awards.

Sample Shimano Recognition Certificate

SHIMANO
Big Fish Honor Roll

This is to certify that
Larry Barker
Has been enrolled in the *Big Fish Honor Roll*
by Shimano American Corporation for catching a
14 inches — Black Crappie
Size — Species
This was one of the top catches of this species reported to Shimano in 2002.

Also, several other fishing clubs, outdoor publications and associations sponsor annual contests for biggest fish during a given period (usually during a calendar year). One such association is the North American Fishing Club. Information about their organization and the annual fishing contest can be obtained by writing to: NAFC, P.O. Box 3403, Minnetonka, Minnesota 55343. You can also get information online at their website: fishingclub.com.

Sample NAFC Recognition Certificate

North American Fishing Club

Larry Barker
for participation in the 2002 Catch & Release Awards program
26 inch Largemouth Bass

Steve Pennaz
Club Director

Chapter Six

Lake Tarpon Resources

Chapter Six

Lake Tarpon Resources

Throughout the book we've referred to various commercial, state and county agencies. We've at times provided web site addresses and/or phone numbers. In this chapter we provide more comprehensive information about resources available to Lake Tarpon anglers. It should be emphasized that, since Lake Tarpon is over eight miles long and highways border it on three sides, there are numerous commercial establishments available to Lake Tarpon anglers. Those identified below are some that are closest to the lake, and that provide services especially useful to anglers.

Lodging/Lake Access Restaurants

The Tarpon Turtle Restaurant & Resort also is accessible from either land or water. Its address is 1513 Lake Tarpon Avenue. Anglers wanting to arrive by boat can see the Tarpon Turtle sign and docks by going north from Anderson Park and watching for the "Turtle" near the northwest end of the lake. You can rent cabins at the "Turtle," or just stop by for a drink or meal. The lake views from the bar and restaurant are outstanding. To order a take out meal call (727) 934-3696, and the Tarpon Turtle will have it waiting for you within 10-15 minutes.

The Sheraton Four Points Hotel and Restaurant is located at **37611 U.S. Hwy 19 N**, but is also accessible via a canal from the lake itself. Reservations can be made by calling (727) **942-0358**.

Marinas/Boat Sales and Rentals

Boat & Motor Superstores is located at **36851 U.S. Hwy 19 N**, and is accessible by canal through Little Dolly Bay. Watch for signs guiding you to the *Boat & Motor Superstores* canal (and remember that all canals on the lake are "no wake" zones.) *Boat & Motor Superstores* is family owned and has been in business since 1962. It sells both freshwater and saltwater boats and a variety of motors, including Mercury, Honda and Yamaha. There is a complete service bay as well as a ship's store on the lot, with accessories and replacement parts for a variety of boats and motors. *Boat & Motor Superstores* phone number is **(727) 942-7767**.

CNR Watercraft Rental is located at **40081 U.S. Hwy 19 N**, and is accessible by canal on the northwest end of Salmon Bay in Anderson Park. *CNR* rents a variety of watercraft to boaters and anglers. Jet Skis and Wave Runners are among the most popular watercraft that CNR rents, but they also rent fishing and paddle boats. The *CNR* phone number is **(727) 937-3933**.

Fischer Marine is located at **37517 Us Hwy 19 N**, and is accessible by canal from the north end of Dolly Bay. The phone number is **(727) 934-0003**.

Bait and Tackle

Tarpon Tom's Bait and Tackle is located at **36725 U.S. Hwy 19 N** (adjacent to Kenyon Boat and Motor Superstore). It boasts the largest supply of fresh and saltwater tackle close to Lake Tarpon, and has a large variety of live and artificial baits. If you are looking for lake shiners to chase that trophy bass, *Tarpon Tom's* is the place to get them. The shiners are of high quality, good size and are "cured" prior to sale.

Tarpon Tom's also keeps lake records, and serves as a clearinghouse for information on local and state tournaments, meetings and special events. *Tarpon Tom's* phone number is **(727) 938-2379**.

Florida Fish and Wildlife Conservation Commission

Web address: **floridafisheries.com**

This web site can link you to a variety of Florida Fisheries regulations and programs. Current creel limits and sizes, unique regulations for Fish Management Areas (including Lake Tarpon), top ten lakes for bass, crappie, and bream, and species identification are just a few of the data bases that the web site hosts.

To obtain a fishing license:

On the web: **floridafisheries.com/license.html**

By phone: **1-888-FISH-FLO (1-888-347-4356)**

Government Contacts

All of the following agencies have some relationship to or jurisdiction over Lake Tarpon:

Pinellas County Department of Environmental Management: **(727) 464-4761**

(to report environmental concerns)

Pinellas County Mosquito Control: **(727) 464-7503**

Florida Fish and Wildlife Conservation Commission: (800) 282-8002

(to report aggressive alligators)

Pinellas County Commissioners: **(727) 464-3377**

Pinellas County Sheriff's Office/ Marine Unit: **(727) 582-6200**

(to report illegal activities on the lake)

Southwest Florida Water Management District: **(800) 423-1476**

Florida Department of Environmental Protection: **(850) 488-1554**

Organizations that Maintain Fishing Records

International Game Fish Association:

Web address: **igfa.org**

Fax number: **(954) 924-4299**

The IGFA is the premier salt- and freshwater fishing record certifying agency. It maintains state and world records for hundreds of different species of fish, and for a variety of line tests for each species. Check the web site for information about joining IGFA and establishing records for your trophy fish.

National Fresh Water Fishing Hall of Fame:

Web site: **freshwater-fishing.org**

Phone Number: **(715) 634-4440**

Established in 1970, the NFWFHF designed a system to qualify and maintain records for most species of fresh water fish. The organization certifies records for different line classes, as well as for catch and release trophy fish. Headquarters in Hayward, Wisconsin, is also home to the Hall of Fame, where memorabilia, photos and information about all records are displayed.

Florida Fish and Wildlife Conservation Commission:

Web site: **floridafisheries.com**

The FFWCC maintains state records for a variety of freshwater and saltwater fish caught in the state of Florida. As previously mentioned in this guide, the FFWCC also has a "big catch" program that offers award certificates to anglers (both adult and youth divisions) who catch fish over minimum qualifying lengths. Check the web site for "big catch" minimum length requirements, current state records, as well as the procedure for establishing new records for trophy fish.

Tarpon Tom's Bait and Tackle:

Web site: **tarpon-toms.com**

Tarpon Tom's, located at **36725 US Hwy 19 N**, certifies and maintains records for most species of trophy fish caught in Lake Tarpon. Stop by the shop or call **(727) 938-2379** to find out more about current lake records and the process to certify your record catch.

Appendices

Appendix A

Lake Tarpon Fishing Record Application

Lake Tarpon Fishing Record Application

Species: ____

Weight: **lbs:**____ **oz:**_____ **Length:**____________________

Date of Catch:____________________ **Time of Catch:**____________

Method of Catch (*live bait, casting, fly fishing, etc.*):____________

Bait/Lure used:____________________ **Line/Tippet Test:**

Rod Make: ______________________ **Reel Make:**

Boat Make (*if used*):______________ **Motor Make** ___________

Angler (*Print name as you wish it to appear on certificates and press releases*):

Name __________________________ **Phone**____________________

Permanent Address: ______________________

____

E-mail address: __________________________

Fishing license number/date of expiration:

WITNESSES

Witness to catch (*list name and address of one witnesses to catch*):

____

____

Number of persons witnessing catch ______

Witness to weigh in (*list name and address of witness, other than angler or weigh master to weigh in*):

____

____

AFFIDAVIT

I, the undersigned, hereby take oath and attest that the fish described in this application was hooked, fought, and landed by me without assistance from any one (*with the exception of help with netting*), and that it was caught in accordance with State of Florida fishing laws and regulations. I further declare that all information in this application is true and correct to the best of my knowledge. I agree to verification procedures and hold harmless, in the event of disputes, the LAKE TARPON FISHING RECORD CERTIFYING AGENCY- - Tarpon-Tom's Bait and Tackle. I give permission to use photographs of me and my catch in press and/or publicity materials regarding my record

Angler signature:____________________

Witness signature:____________________

Date: ____________

I, the weigh master for the LAKE TARPON FISHING RECORD CERTIFYING AGENCYTarpon-Tom's Bait and Tackle, hereby certify that the weight listed on this applcation is true and correct and that the scales used have been certified by ____________________.

Weigh master signature:____________________

Date: ____________

Appendix B

Tarpon Tom's Annual Fishing Contest

Press Release

"Tarpon-Toms Sponsors New Lake Tarpon Fishing Contest"

Capt. Lenny Crispino, owner of Tarpon-Tom's Bait and Tackle recently announced that his shop will sponsor a monthly and annual fishing contest on Lake Tarpon. The annual contest will begin on January 1 each year and end on December 31. Anglers catching the heaviest fish in each eligible species during the year will receive certificates, pins and gift certificates. Monthly Award Certificates will also be given for the heaviest fish entered in each of the 3 eligible species. The species include largemouth bass, bluegill (including shellcrackers and sunfish) and Specks (white or black crappie).

In addition, Tarpon Toms will begin keeping official Lake Tarpon Fish Records for the above species. In the past, only unofficial records have been kept, and these have not been recognized or accepted by such agencies as the International Game Fish Association, the National Freshwater Hall of Fame, or the Florida Fish and Wildlife Conservation Commission. To help local anglers certify record catches for all agencies, Capt. Crispino has installed a state certified scale for anglers to weigh in their prize catches.

Anglers can get rules and entry forms on the web at: tarpon-toms.com, or they can pick them up at the shop (36725 US Hwy 19 N., Palm Harbor: Phone: 727-938-2379).

TARPON-TOM'S ANNUAL FISHING CONTEST

Rules:

*Fish entered in the contest must be caught in Lake Tarpon (including Brooker Creek and the Lake Tarpon Outfall Canal)

*All entries must be brought to Tarpon-Tom's Bait and Tackle shop during regular business hours within 12 hours of catch for weigh in and certification. All weigh-ins must be witnessed by a person other than the angler or weigh master.

*Entries must include a photograph that meets the following requirements:

- Fish must be held in a manner in photo that doesn't harm fish
- Photo must show the angler on or near the water
- Print angler's name, address and phone number on the back of photo
- 35 mm, Polaroid, and digital camera disks/photos are acceptable.

*All fish must be hooked, fought, and landed by the angler (*with the exception of any help needed netting the fish) in accordance* with the State of Florida fishing laws and regulations.

*Angler agrees to verification procedures, if required, and certifies that he/she will hold Tarpon-Tom's Bait and Tackle harmless in the event of disputes.

*All decisions of the Contest Chairman will be final.

*All largemouth bass entered in the contest must be released within 2 hours of weigh in.

Eligible Species:

- **Largemouth Bass**
- **Bluegill (including shellcrackers and sunfish)**
- **Specks (White or Black Crappie)**

Awards:

**Monthly Award Certificates* will be given for the heaviest fish caught and entered for each eligible species. <u>*(Monthly contests begin at 12:01 AM on the first day of each month and end at 11:59 PM on the last day of the month)*</u>

*Monthly winners for each eligible species automatically will be entered in the yearly contest.

**Yearly Award Certificates,* and gift certificates will be given for the heaviest fish caught and entered for each eligible species. <u>*(Yearly contests begin at 12:01 AM on January 1, and ends on Midnight, December 31, each year)*</u>

*Additional merchandise and prizes, provided by contest sponsors, may also be given for outstanding monthly and yearly catches, at the discretion of the contest chairman.

*All leading entries for each species will be posted on the web site: tarpontoms.com. Monthly and annual winners for each eligible species will be posted on the same web site. Anglers wanting to know the status of their entry may also call 727-938-2379.

*All winning anglers must pick up their awards at Tarpon Tom's Bait and Tackle within 30 days after being certified as a winner in either the monthly or yearly contest.

Appendix C
Reproducible Fishing Logs

(Courtesy of BassResource.com)

FISHING LOG

Location: ______________________ Date:______ Day of Week: __________

County: ____________________ Time Started:______ Time Finished:__________

Water Temp: Start:________ End:_________ Air Temp: Start:________ End:_______

STRUCTURE TYPES:
☐ Docks ☐ Stumps ☐ Brush/Trees ☐ Rocks ☐ Rip-Rap ☐ Hump ☐ Rock Piles

☐ Points ☐ Road Beds ☐ Rock Cliffs ☐ Mud Flats ☐ Gravel Banks ☐ Flooded Timber

Other______________ Other ______________

AQUATIC VEGETATION:
☐ Lily Pads ☐ Cattails ☐ Milfoil ☐ Dollar Pads ☐ Bulrushes ☐ Elodea ☐ Coontail

☐ Maidencane ☐ Cabbage ☐ Hydrilla Other __________ Other ______________

Growth Stage: ☐ Submerged ☐ Emergent

WEATHER:
☐ Bright Sun ☐ Partly Cloudy ☐ Overcast ☐ Showers ☐ Rain Other______________

Wind MPH:_______ DIRECTION ______

WATER CLARITY: ☐ Clear ☐ Stained ☐ Murky ☐ Muddy Other____________

WATER LEVEL: ☐ Normal ☐ High ☐ Low ☐ Rising ☐ Dropping

CONTRIBUTING FACTORS: ☐ Baitfish ☐ Insects ☐ Crayfish Other______________

TOTAL NUMBER OF FISH: ____________

FISH OVER TWELVE INCHES:

	L/M	S/M	LENGTH	WEIGHT	TIME	LURE TYPE/COLOR/WEIGHT/NAME
1.						
2.						
3.						
4.						
5.						
6.						
7.						
8.						
9.						
10.						

NOTES:

MOON PHASE: ☐ 1st Quarter ☐ ½ Moon ☐ 3rd Quarter ☐ Full Moon

+ ___ Days + ___ Days + ___ Days + ___ Days

FISHING PARTNER: ______________ PAGE ___ OF___

Courtesy BassResource.com

Courtesy BassResource.com

FISHING LOG

Location: ______________________ Date:______ Day of Week: __________

County: ______________________ Time Started:______ Time Finished:__________

Water Temp: Start:________ End:________ Air Temp: Start:________ End:________

STRUCTURE TYPES:
☐ Docks ☐ Stumps ☐ Brush/Trees ☐ Rocks ☐ Rip-Rap ☐ Hump ☐ Rock Piles

☐ Points ☐ Road Beds ☐ Rock Cliffs ☐ Mud Flats ☐ Gravel Banks ☐ Flooded Timber

Other______________ Other ______________

AQUATIC VEGETATION:
☐ Lily Pads ☐ Cattails ☐ Milfoil ☐ Dollar Pads ☐ Bulrushes ☐ Elodea ☐ Coontail

☐ Maidencane ☐ Cabbage ☐ Hydrilla Other ____________ Other ____________

Growth Stage: ☐ Submerged ☐ Emergent

WEATHER:
☐ Bright Sun ☐ Partly Cloudy ☐ Overcast ☐ Showers ☐ Rain Other__________

Wind MPH:_______ DIRECTION ______

WATER CLARITY: ☐ Clear ☐ Stained ☐ Murky ☐ Muddy Other__________

WATER LEVEL: ☐ Normal ☐ High ☐ Low ☐ Rising ☐ Dropping

CONTRIBUTING FACTORS: ☐ Baitfish ☐ Insects ☐ Crayfish Other__________

TOTAL NUMBER OF FISH: ____________

FISH OVER TWELVE INCHES:

	L/M	S/M	LENGTH	WEIGHT	TIME	LURE TYPE/COLOR/WEIGHT/NAME
1.						
2.						
3.						
4.						
5.						
6.						
7.						
8.						
9.						
10						

NOTES:

MOON PHASE: ☐ 1st Quarter ☐ ½ Moon ☐ 3rd Quarter ☐ Full Moon

+___ Days +___ Days +___ Days +___ Days

FISHING PARTNER:______________ PAGE ___OF___

Courtesy BassResource.com

FISHING LOG

Location: ____________ Date:______ Day of Week: ____________

County: ____________ Time Started:______ Time Finished:____________

Water Temp: Start:________ End:________ Air Temp: Start:________ End:________

STRUCTURE TYPES:
☐ Docks ☐ Stumps ☐ Brush/Trees ☐ Rocks ☐ Rip-Rap ☐ Hump ☐ Rock Piles

☐ Points ☐ Road Beds ☐ Rock Cliffs ☐ Mud Flats ☐ Gravel Banks ☐ Flooded Timber

Other____________ Other ____________

AQUATIC VEGETATION:
☐ Lily Pads ☐ Cattails ☐ Milfoil ☐ Dollar Pads ☐ Bulrushes ☐ Elodea ☐ Coontail

☐ Maidencane ☐ Cabbage ☐ Hydrilla Other ____________ Other ____________

Growth Stage: ☐ Submerged ☐ Emergent

WEATHER:
☐ Bright Sun ☐ Partly Cloudy ☐ Overcast ☐ Showers ☐ Rain Other____________

Wind MPH:______ DIRECTION ______

WATER CLARITY: ☐ Clear ☐ Stained ☐ Murky ☐ Muddy Other____________

WATER LEVEL: ☐ Normal ☐ High ☐ Low ☐ Rising ☐ Dropping

CONTRIBUTING FACTORS: ☐ Baitfish ☐ Insects ☐ Crayfish Other____________

TOTAL NUMBER OF FISH: ____________

FISH OVER TWELVE INCHES:

	L/M	S/M	LENGTH	WEIGHT	TIME	LURE TYPE/COLOR/WEIGHT/NAME
1.						
2.						
3.						
4.						
5.						
6.						
7.						
8.						
9.						
10						

NOTES:

MOON PHASE: ☐ 1st Quarter ☐ ½ Moon ☐ 3rd Quarter ☐ Full Moon

+ ___ Days + ___ Days + ___ Days + ___ Days

FISHING PARTNER: ____________ PAGE ___ OF ___

Courtesy BassResource.com

FISHING LOG

Location: ______________________ Date:______ Day of Week: __________

County: ______________________ Time Started:______ Time Finished:__________

Water Temp: Start:________ End:__________ Air Temp: Start:_________ End:_______

STRUCTURE TYPES:
☐ Docks ☐ Stumps ☐ Brush/Trees ☐ Rocks ☐ Rip-Rap ☐ Hump ☐ Rock Piles

☐ Points ☐ Road Beds ☐ Rock Cliffs ☐ Mud Flats ☐ Gravel Banks ☐ Flooded Timber

Other_____________ Other _____________

AQUATIC VEGETATION:
☐ Lily Pads ☐ Cattails ☐ Milfoil ☐ Dollar Pads ☐ Bulrushes ☐ Elodea ☐ Coontail

☐ Maidencane ☐ Cabbage ☐ Hydrilla Other __________ Other ______________

Growth Stage: ☐ Submerged ☐ Emergent

WEATHER:
☐ Bright Sun ☐ Partly Cloudy ☐ Overcast ☐ Showers ☐ Rain Other______________

Wind MPH:_______ DIRECTION ______

WATER CLARITY: ☐ Clear ☐ Stained ☐ Murky ☐ Muddy Other____________

WATER LEVEL: ☐ Normal ☐ High ☐ Low ☐ Rising ☐ Dropping

CONTRIBUTING FACTORS: ☐ Baitfish ☐ Insects ☐ Crayfish Other______________

TOTAL NUMBER OF FISH: ____________

FISH OVER TWELVE INCHES:

	L/M	S/M	LENGTH	WEIGHT	TIME	LURE TYPE/COLOR/WEIGHT/NAME
1.						
2.						
3.						
4.						
5.						
6.						
7.						
8.						
9.						
10						

NOTES:

MOON PHASE: ☐ 1st Quarter ☐ ½ Moon ☐ 3rd Quarter ☐ Full Moon

+___ Days +___ Days +___ Days +___ Days

FISHING PARTNER:______________ PAGE ___OF___

Courtesy BassResource.com

FISHING LOG

Location: ______ Date:______ Day of Week: ______

County: ______ Time Started:______ Time Finished:______

Water Temp: Start:______ End:______ Air Temp: Start:______ End:______

STRUCTURE TYPES:
☐ Docks ☐ Stumps ☐ Brush/Trees ☐ Rocks ☐ Rip-Rap ☐ Hump ☐ Rock Piles

☐ Points ☐ Road Beds ☐ Rock Cliffs ☐ Mud Flats ☐ Gravel Banks ☐ Flooded Timber

Other______ Other ______

AQUATIC VEGETATION:
☐ Lily Pads ☐ Cattails ☐ Milfoil ☐ Dollar Pads ☐ Bulrushes ☐ Elodea ☐ Coontail

☐ Maidencane ☐ Cabbage ☐ Hydrilla Other ______ Other ______

Growth Stage: ☐ Submerged ☐ Emergent

WEATHER:
☐ Bright Sun ☐ Partly Cloudy ☐ Overcast ☐ Showers ☐ Rain Other______

Wind MPH:______ DIRECTION ______

WATER CLARITY: ☐ Clear ☐ Stained ☐ Murky ☐ Muddy Other______

WATER LEVEL: ☐ Normal ☐ High ☐ Low ☐ Rising ☐ Dropping

CONTRIBUTING FACTORS: ☐ Baitfish ☐ Insects ☐ Crayfish Other______

TOTAL NUMBER OF FISH: ______

FISH OVER TWELVE INCHES:

	L/M	S/M	LENGTH	WEIGHT	TIME	LURE TYPE/COLOR/WEIGHT/NAME
1.						
2.						
3.						
4.						
5.						
6.						
7.						
8.						
9.						
10						

NOTES: ______

______

______

MOON PHASE: ☐ 1st Quarter ☐ ½ Moon ☐ 3rd Quarter ☐ Full Moon

+ ___ Days + ___ Days + ___ Days + ___ Days

FISHING PARTNER: ______ PAGE ___ OF ___

FISHING LOG

Location: ______________ Date:______ Day of Week: ____________

County: ______________ Time Started:______ Time Finished:____________

Water Temp: Start:________ End:_________ Air Temp: Start:_________ End:________

STRUCTURE TYPES:
☐ Docks ☐ Stumps ☐ Brush/Trees ☐ Rocks ☐ Rip-Rap ☐ Hump ☐ Rock Piles

☐ Points ☐ Road Beds ☐ Rock Cliffs ☐ Mud Flats ☐ Gravel Banks ☐ Flooded Timber

Other______________ Other ______________

AQUATIC VEGETATION:
☐ Lily Pads ☐ Cattails ☐ Milfoil ☐ Dollar Pads ☐ Bulrushes ☐ Elodea ☐ Coontail

☐ Maidencane ☐ Cabbage ☐ Hydrilla Other ____________ Other ______________

Growth Stage: ☐ Submerged ☐ Emergent

WEATHER:
☐ Bright Sun ☐ Partly Cloudy ☐ Overcast ☐ Showers ☐ Rain Other______________

Wind MPH:_______ DIRECTION ______

WATER CLARITY: ☐ Clear ☐ Stained ☐ Murky ☐ Muddy Other______________

WATER LEVEL: ☐ Normal ☐ High ☐ Low ☐ Rising ☐ Dropping

CONTRIBUTING FACTORS: ☐ Baitfish ☐ Insects ☐ Crayfish Other______________

TOTAL NUMBER OF FISH: ____________

FISH OVER TWELVE INCHES:

	L/M	S/M	LENGTH	WEIGHT	TIME	LURE TYPE/COLOR/WEIGHT/NAME
1.						
2.						
3.						
4.						
5.						
6.						
7.						
8.						
9.						
10.						

NOTES:

MOON PHASE: ☐ 1st Quarter ☐ ½ Moon ☐ 3rd Quarter ☐ Full Moon

+ ___ Days + ___ Days + ___ Days + ___ Days

FISHING PARTNER: ______________ PAGE ___ OF___

FISHING LOG

Location: ______________________ Date:______ Day of Week: __________

County: ____________________ Time Started:______ Time Finished:__________

Water Temp: Start:________ End: __________ Air Temp: Start: _________ End:________

STRUCTURE TYPES:
☐ Docks ☐ Stumps ☐ Brush/Trees ☐ Rocks ☐ Rip-Rap ☐ Hump ☐ Rock Piles

☐ Points ☐ Road Beds ☐ Rock Cliffs ☐ Mud Flats ☐ Gravel Banks ☐ Flooded Timber

Other______________ Other ______________

AQUATIC VEGETATION:
☐ Lily Pads ☐ Cattails ☐ Milfoil ☐ Dollar Pads ☐ Bulrushes ☐ Elodea ☐ Coontail

☐ Maidencane ☐ Cabbage ☐ Hydrilla Other ___________ Other ______________

Growth Stage: ☐ Submerged ☐ Emergent

WEATHER:
☐ Bright Sun ☐ Partly Cloudy ☐ Overcast ☐ Showers ☐ Rain Other______________

Wind MPH:_______ DIRECTION ______

WATER CLARITY: ☐ Clear ☐ Stained ☐ Murky ☐ Muddy Other____________

WATER LEVEL: ☐ Normal ☐ High ☐ Low ☐ Rising ☐ Dropping

CONTRIBUTING FACTORS: ☐ Baitfish ☐ Insects ☐ Crayfish Other______________

TOTAL NUMBER OF FISH: ____________

FISH OVER TWELVE INCHES:

	L/M	S/M	LENGTH	WEIGHT	TIME	LURE TYPE/COLOR/WEIGHT/NAME
1.						
2.						
3.						
4.						
5.						
6.						
7.						
8.						
9.						
10						

NOTES: ____

____

____

MOONPHASE: ☐ 1st Quarter ☐ ½ Moon ☐ 3rd Quarter ☐ Full Moon

+____ Days +____ Days +____ Days +____ Days

FISHING PARTNER:______________ PAGE ____OF____

FISHING LOG

Location: ______________________ Date:______ Day of Week: __________

County: ____________________ Time Started:______ Time Finished:__________

Water Temp: Start:________ End:__________ Air Temp: Start:________ End:________

STRUCTURE TYPES:
☐ Docks ☐ Stumps ☐ Brush/Trees ☐ Rocks ☐ Rip-Rap ☐ Hump ☐ Rock Piles

☐ Points ☐ Road Beds ☐ Rock Cliffs ☐ Mud Flats ☐ Gravel Banks ☐ Flooded Timber

Other______________ Other ______________

AQUATIC VEGETATION:
☐ Lily Pads ☐ Cattails ☐ Milfoil ☐ Dollar Pads ☐ Bulrushes ☐ Elodea ☐ Coontail

☐ Maidencane ☐ Cabbage ☐ Hydrilla Other __________ Other ______________

Growth Stage: ☐ Submerged ☐ Emergent

WEATHER:
☐ Bright Sun ☐ Partly Cloudy ☐ Overcast ☐ Showers ☐ Rain Other______________

Wind MPH:_______ DIRECTION ______

WATER CLARITY: ☐ Clear ☐ Stained ☐ Murky ☐ Muddy Other____________

WATER LEVEL: ☐ Normal ☐ High ☐ Low ☐ Rising ☐ Dropping

CONTRIBUTING FACTORS: ☐ Baitfish ☐ Insects ☐ Crayfish Other______________

TOTAL NUMBER OF FISH: ____________

FISH OVER TWELVE INCHES:

	L/M	S/M	LENGTH	WEIGHT	TIME	LURE TYPE/COLOR/WEIGHT/NAME
1.						
2.						
3.						
4.						
5.						
6.						
7.						
8.						
9.						
10						

NOTES:

____

____

____

MOON PHASE: ☐ 1st Quarter ☐ ½ Moon ☐ 3rd Quarter ☐ Full Moon

+ ____ Days + ____ Days + ____ Days + ____ Days

FISHING PARTNER: ______________ PAGE ____ OF ____

Courtesy BassResource.com

Courtesy BassResource.com

FISHING LOG

Location: ______________________________ Date:______ Day of Week: ____________

County: ________________________ Time Started:______ Time Finished:____________

Water Temp: Start:________ End:__________ Air Temp: Start:_________ End:________

STRUCTURE TYPES:
☐ Docks ☐ Stumps ☐ Brush/Trees ☐ Rocks ☐ Rip-Rap ☐ Hump ☐ Rock Piles

☐ Points ☐ Road Beds ☐ Rock Cliffs ☐ Mud Flats ☐ Gravel Banks ☐ Flooded Timber

Other______________ Other ______________

AQUATIC VEGETATION:
☐ Lily Pads ☐ Cattails ☐ Milfoil ☐ Dollar Pads ☐ Bulrushes ☐ Elodea ☐ Coontail

☐ Maidencane ☐ Cabbage ☐ Hydrilla Other ____________ Other ______________

Growth Stage: ☐ Submerged ☐ Emergent

WEATHER:
☐ Bright Sun ☐ Partly Cloudy ☐ Overcast ☐ Showers ☐ Rain Other______________

Wind MPH:_______ DIRECTION ______

WATER CLARITY: ☐ Clear ☐ Stained ☐ Murky ☐ Muddy Other______________

WATER LEVEL: ☐ Normal ☐ High ☐ Low ☐ Rising ☐ Dropping

CONTRIBUTING FACTORS: ☐ Baitfish ☐ Insects ☐ Crayfish Other________________

TOTAL NUMBER OF FISH: ____________

FISH OVER TWELVE INCHES:

	L/M	S/M	LENGTH	WEIGHT	TIME	LURE TYPE/COLOR/WEIGHT/NAME
1.						
2.						
3.						
4.						
5.						
6.						
7.						
8.						
9.						
10						

NOTES:

MOON PHASE: ☐ 1st Quarter ☐ ½ Moon ☐ 3rd Quarter ☐ Full Moon

+____ Days +____ Days +____ Days +____ Days

FISHING PARTNER:________________ PAGE ____OF____

Courtesy BassResource.com

FISHING LOG

Location: ____________________ Date:______ Day of Week: __________

County: ____________________ Time Started:______ Time Finished:__________

Water Temp: Start:________ End: ________ Air Temp: Start: ________ End:________

STRUCTURE TYPES:
☐ Docks ☐ Stumps ☐ Brush/Trees ☐ Rocks ☐ Rip-Rap ☐ Hump ☐ Rock Piles

☐ Points ☐ Road Beds ☐ Rock Cliffs ☐ Mud Flats ☐ Gravel Banks ☐ Flooded Timber

Other______________ Other ______________

AQUATIC VEGETATION:
☐ Lily Pads ☐ Cattails ☐ Milfoil ☐ Dollar Pads ☐ Bulrushes ☐ Elodea ☐ Coontail

☐ Maidencane ☐ Cabbage ☐ Hydrilla Other __________ Other ______________

Growth Stage: ☐ Submerged ☐ Emergent

WEATHER:
☐ Bright Sun ☐ Partly Cloudy ☐ Overcast ☐ Showers ☐ Rain Other______________

Wind MPH:_______ DIRECTION ______

WATER CLARITY: ☐ Clear ☐ Stained ☐ Murky ☐ Muddy Other____________

WATER LEVEL: ☐ Normal ☐ High ☐ Low ☐ Rising ☐ Dropping

CONTRIBUTING FACTORS: ☐ Baitfish ☐ Insects ☐ Crayfish Other______________

TOTAL NUMBER OF FISH: ____________

FISH OVER TWELVE INCHES:

	L/M	S/M	LENGTH	WEIGHT	TIME	LURE TYPE/COLOR/WEIGHT/NAME
1.						
2.						
3.						
4.						
5.						
6.						
7.						
8.						
9.						
10.						

NOTES:

MOON PHASE: ☐ 1st Quarter ☐ ½ Moon ☐ 3rd Quarter ☐ Full Moon

+___ Days +___ Days +___ Days +___ Days

FISHING PARTNER:______________ PAGE ___OF___

FISHING LOG

Location: ______________________ Date:______ Day of Week: __________

County: ____________________ Time Started:______ Time Finished:__________

Water Temp: Start:________ End:__________ Air Temp: Start:________ End:________

STRUCTURE TYPES:
☐ Docks ☐ Stumps ☐ Brush/Trees ☐ Rocks ☐ Rip-Rap ☐ Hump ☐ Rock Piles

☐ Points ☐ Road Beds ☐ Rock Cliffs ☐ Mud Flats ☐ Gravel Banks ☐ Flooded Timber

Other______________ Other ______________

AQUATIC VEGETATION:
☐ Lily Pads ☐ Cattails ☐ Milfoil ☐ Dollar Pads ☐ Bulrushes ☐ Elodea ☐ Coontail

☐ Maidencane ☐ Cabbage ☐ Hydrilla Other __________ Other ______________

Growth Stage: ☐ Submerged ☐ Emergent

WEATHER:
☐ Bright Sun ☐ Partly Cloudy ☐ Overcast ☐ Showers ☐ Rain Other__________

Wind MPH:______ DIRECTION ______

WATER CLARITY: ☐ Clear ☐ Stained ☐ Murky ☐ Muddy Other__________

WATER LEVEL: ☐ Normal ☐ High ☐ Low ☐ Rising ☐ Dropping

CONTRIBUTING FACTORS: ☐ Baitfish ☐ Insects ☐ Crayfish Other__________

TOTAL NUMBER OF FISH: __________

FISH OVER TWELVE INCHES:

	L/M	S/M	LENGTH	WEIGHT	TIME	LURE TYPE/COLOR/WEIGHT/NAME
1.						
2.						
3.						
4.						
5.						
6.						
7.						
8.						
9.						
10						

NOTES:

MOON PHASE: ☐ 1st Quarter ☐ ½ Moon ☐ 3rd Quarter ☐ Full Moon

+ ___ Days + ___ Days + ___ Days + ___ Days

FISHING PARTNER: ______________ PAGE ___ OF ___

Courtesy BassResource.com

About our Contributing Editor

Dan Richardson

Dan Richardson, Vice President of Wachovia Bank, is a veteran tournament fisherman. He has been competing in bass tournaments for over 25 years, and has over 140 trophies and awards to show for his efforts. He fished the Ranger owner's tournament recently with nearly 150 anglers and finished in the top 20. Dan has fished Lake Tarpon most of his life, and has been a member of local bass clubs since age 14. His grasp of spinnerbaits and crankbaits is unsurpassed— and he is an excellent fishing coach/instructor. He is an outstanding speaker and frequently gives seminars and talks to social, civic and Bass clubs about freshwater fishing.

Notes

ISBN 1553952448
9 781553 952442